OHIO ROCKS!

A Guide to Geologic Sites in the Buckeye State

ALBERT BINKLEY DICKAS

OHIO ROCKS!

A Guide to Geologic Sites in the Buckeye State

ALBERT BINKLEY DICKAS

2015
Mountain Press Publishing Company
Missoula, Montana

Third Printing, May 2019

Cover image: Rock House. *—Courtesy of the Ohio Department of Natural Resources, Division of Geological Survey*

Library of Congress Cataloging-in-Publication Data

Dickas, Albert Binkley, 1933-
Ohio rocks! : a guide to geologic sites in the Buckeye State / Albert Binkley Dickas.
pages cm
Includes bibliographical references and index.
ISBN 978-0-87842-635-5 (pbk. : alk. paper)
1. Geology—Ohio—Guidebooks. 2. Ohio—Guidebooks. I. Title.
QE151.D53 2014
557.71—dc23
2014031321

PRINTED IN THE USA BY VERSA PRESS, INC.

MP Mountain Press
PUBLISHING COMPANY
P.O. Box 2399 • Missoula, MT 59806 • 406-728-1900
800-234-5308 • info@mtnpress.com
www.mountain-press.com

In memory of my father, who early on encouraged me
to appreciate and respect the many facets of the world of nature.

PREFACE

Throughout much of the twentieth century, mainstream America knew little of the science of geology. For decades most geologists not engaged in education had been involved in the exploration for oil, gas, coal, or ore deposits—investigations that had long been somewhat static in terms of scientific innovation and informative publicity. Then came change.

In the mid-1960s, the universal acceptance of the theory of plate tectonics transformed the world of science. The theory posits that the worldwide distribution of geological phenomena, such as mountain building, volcanism, and seismicity, is directly related to the formation, movement, and destruction of a series of rigid plates that interlock together to make up Earth's lithosphere. The theory was at first questioned, but when it gained credence its effect was as cataclysmic as any of those forces themselves.

In 1970, the celebration of the first Earth Day gave the American public an effective means to express its concern about the perceived deterioration of Earth's land, water, and atmosphere. Overnight the word *environment* became a subject for discussion within a wide range of venues, from the evening news to the halls of government.

As a result of these two watershed declarations—plate tectonics and Earth Day—geology, the study of planet Earth, today occupies a leadership role in the annals of science, and the public at large is much more informed about the wonders and earthly manipulations of Mother Nature and her ability to create beauty and order over the long term and death and destruction in the short term. For many reasons we are more acutely aware of how the dictates of nature affect humanity: print and electronic programming focusing on the history of Earth; our understanding of the causes of earthquakes, tsunamis, and volcanic eruptions; and our growing knowledge about the effects of global warming and the melting of the Antarctic ice sheets.

Ohio is blessed with a variety of locales closely associated with interesting and educational geologic topics: glaciation, coal formation, extraterrestrial impact, caverns, oil and natural gas fields, fossil sites, Great Lakes formation, and even a ghost river that once connected the western slopes of the Appalachian Mountains with the ancestral Mississippi River. All of these topics, and more, are presented herein in text reflective of new-era concepts of geologic understanding and interpretation.

The majority of the fifty geologic sites discussed in this travel guide are located on protected, publicly accessible city, state, and federal land. For safety reasons, the few rock exposures that are situated along the side of major highways should be considered solely as drive-by locations. Please do not trespass or collect rock or fossil specimens anywhere unless permission is first obtained. For the sites that operate on a schedule, be sure to check websites or call ahead before departure to determine seasons, days, and hours of operation.

It is my wish that your visits to these unusual locales will enhance your comprehension of the processes through which they formed, and that you will gain a better understanding of how their presence has defined the topographic profile of the state of Ohio. Maybe we will run into each other at one site or another. I hope so.

Albert Binkley Dickas
Brush Mountain, Virginia
September 2014

ACKNOWLEDGMENTS

The writing of a field guide such as this, highlighting locales of unusual geology, cannot be accomplished in isolation. Many individuals are involved, and their help, suggestions, and cooperation are hereby gratefully acknowledged.

E. Mac Swinford, now retired from his position as chief of the Division of Geological Survey, Ohio Department of Natural Resources, readily granted me access to the knowledge he gained through numerous years of experience, and to the many personnel and documentary resources of the survey. Two individuals were of special help: geologist Mark T. Baranoski for his thorough review of the text with the exactitude of the professional mind, and Lisa Van Doren, electronic designer in the Publications and Geologic Records Center, for her help in identifying the best of the survey's many outstanding photographs for use in this guide.

My editor at Mountain Press, James Lainsbury, has been, as always, excessive in patience, generous in encouragement, and thorough in editing. I marvel at his continuing ability to mold a higher degree of consistency and understanding in my use of dates, facts, and concepts.

Published author and valued and cherished friend Rachael M. Garrity, principal of Penworthy LLC, tirelessly responded to my frequent, and far too often late-hour requests for assistance in matters of grammar, usage, and syntax. She remains the very soul of tolerance, and her understanding and guidance have been indispensable during those many occasions when my flights of compositional fancy made little sense. My debt to her can never be repaid, but I shall try.

Finally, I would be remiss should I neglect to recognize the many individuals associated with universities and colleges; community, state, and federal parks; and local historical societies and public libraries who helped with the writing of this guide by their suggestions and willingness to share information—their count is well over one hundred. I hope I extended a proper sense of appreciation to each at the time I visited with them, but if I failed, then here and now I say a heartfelt "thank you."

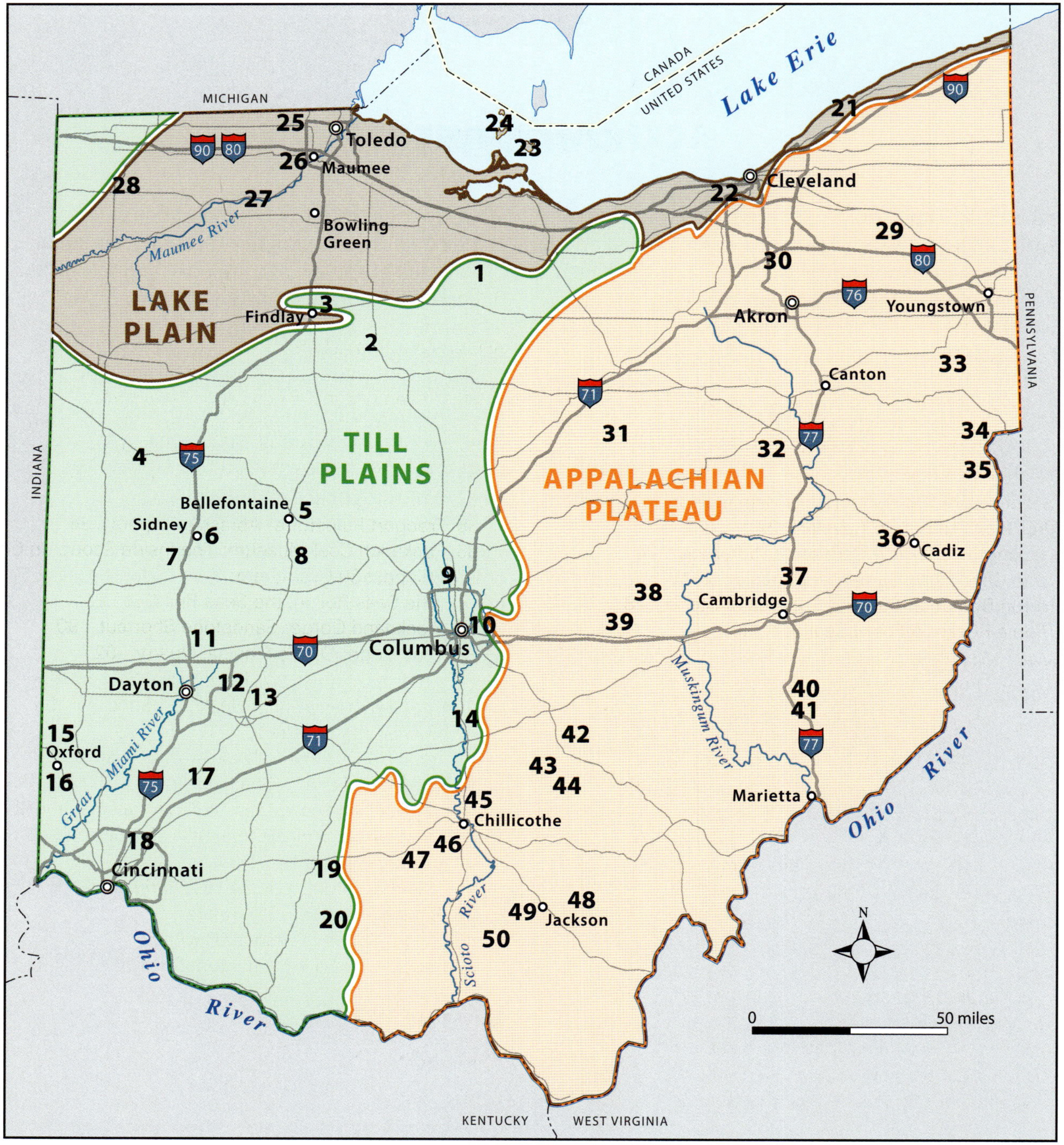

Regions and sites covered in Ohio Rocks! Numbers correspond to the described localities.

CONTENTS

Era	Period	Epoch	Age (millions of years)	Comments
CENOZOIC	**Quaternary**	Holocene	0.01	post ice ages
		Pleistocene	2.6	Ohio's ice ages
	Neogene		23	No rocks or fossils of these ages are found anywhere in Ohio
	Paleogene		66	
MESOZOIC	Cretaceous		145	
	Jurassic		201	
	Triassic		252	
PALEOZOIC	**Permian**		299	The geologic history of Ohio is composed almost entirely of the seven periods that make up the Paleozoic Era
	Pennsylvanian		323	
	Mississippian		359	
	Devonian		419	
	Silurian		444	
	Ordovician		485	
	Cambrian		541	
PRECAMBRIAN			4,540	Basement rocks known only by deep drilling

The geologic timescale. The units of time represented by rocks found in Ohio and discussed in this guide are shown in bold print. Ages are given in units of millions of years.

OHIO'S GEOLOGIC HERITAGE: A NARRATIVE OF THREE EPISODES

I. Precambrian Era: The Beginning

Earth, and the area today designated as the State of Ohio, came into existence approximately 4.54 billion years ago, give or take an uncertainty of 50 million years. This birth date has been determined through the age dating of hundreds of meteorites collected on all seven continents and rocks collected by the American astronauts who walked across the surface of the moon. It is an accepted fact that the planets and moons of our solar system were created at the same general time, thus the ages of extraterrestrial material can be directly related to the age of Earth.

The oldest Earth rock found to date, a windswept exposure lying along the eastern shore of Hudson Bay, is between 3.8 and 4.3 billion years in age. Older rocks may eventually be found, but in all probability normal geologic processes destroyed them long ago, or they lie at depths too great for recovery. In the final analysis, and by any measure of time, one fact stands out with great clarity: Earth is most certainly a senior citizen.

During the early years, Earth was a roiling, boiling sphere of molten rock that was gradually being rearranged into three zones. Heavy elements sank to the center to form a predominantly iron-rich core, and frothy, low-density material rose to the surface to eventually cool as the crust. Material caught in between became the plastic, mushy material of what's called the mantle. A primitive atmosphere enveloped the surface, and as temperatures declined, moisture-laden clouds condensed to produce the first storms. Oceans began to fill in low-lying topography, river systems came into being, and the water cycle was born, along with the ever-present processes of weathering and erosion.

As soon as mobile, Rhode Island–sized landmasses began to crystallize on Earth's surface, the force of upwelling thermal currents, heated by the sun-like temperatures of Earth's core, forced them into lateral movement. When these protocontinents increased in sufficient number, they coalesced to create the rockbound interiors of the present-day continents. Soon, the entire gamut of geologic processes was under way: volcanoes erupted molten spittle, earthquakes haphazardly fractured and rearranged the terrestrial architecture, and mountain ranges rose to lofty heights only to bow to erosion over time, reduced in size to sand and dust. During these early stages, Earth was a seething, rambunctious, and undisciplined planet.

Life, the animated, indispensable state of vitality and being, appeared next on the geologic stage. The biologic curtain rose on a form of single-celled blue-green algae that lacked a true nucleus and reproduced the old-fashioned way—asexually. The photosynthesizing life cycle of these primitive organisms brought about a significant transition in Earth history: the early, toxic, oxygen-starved atmosphere slowly morphed into an inviting, oxygen-rich environment.

By middle Precambrian time, some 2.5 billion years ago, the dynamic processes of plate tectonics, whereby all the continents periodically come together as a giant supercontinent and then are torn apart while oceans are born and then die, were in full swing. This on-average 500-million-year-long cycle of major landmass construction and destruction was responsible for the assembly of Rodinia, a supercontinent that formed some 1.1 billion years ago. It was created by the accretion of the protocontinents of South America, eastern North America, and western Africa.

During the early stages of this continental gathering, Ohio occupied a coastal position, but then, some 900 million years ago, Rodinia collided with a smaller landmass, resulting in the rearrangement of the Buckeye State, sending it to an interior position. The finale of this fender bender was the creation of the Grenville Mountains, a Himalayan-high trend of folded and faulted rock extending from Labrador to Texas.

At the same time, rising levels of atmospheric oxygen sent a signal to the simple asexual life-forms then in existence: evolve or die. Many did expire, but some, fortunately, adapted, evolving to advanced forms that were larger in size, sported a distinct nucleus, and exchanged genes so that each generation possessed a different genetic configuration and biologic variation. In short, the sexual revolution was underway, and Earth began to display its modern complexion.

Except for sporadic glacial erratics, rocks of Precambrian age are nowhere present at the surface in Ohio, but they do underlie the state. Magnetic and gravity analyses and the study of earthquake and artificially generated energy waves have been useful in determining many aspects of their physical makeup and depth. The cartographic version of ancient Ohio that results from this research depicts a buried terrain subdivided by the Grenville Front, a 30-mile-wide, fractured, and highly disturbed zone that marks the western limits of the Grenville Mountains. West of the front 1.5-billion-year-old granite intermixed with sporadic occurrences of 1-billion-year-old sandstone forms the basement, while to the east 800-to-900-million-year-old metamorphic rocks compose the terrain of the Grenville Mountains.

Ohio 541 million years ago, at the end of the Precambrian Era. The terrain west of the Grenville Front was composed of granite and sandstone; to the east the Grenville Mountains were exposed.

An Interim Period

After the formation and uplift of the Grenville Mountains, the entire state of Ohio experienced a long period of erosion that lasted for almost 500 million years. Highland terrain that had once reached beyond the clouds became a landscape of gentle hills, and eventually even the metamorphic roots of the Grenville Mountains were exposed to the corrosive effects of wind and water.

As the Precambrian Era came to an end, the scarred and aged bedrock surface of Ohio could best be described as geologically complex and topographically simple. A long period of planetary birth, growth, evolution, and adjustment had come to an end, and the stage was set for the next chapter of Earth history: the Paleozoic Era.

II. Paleozoic Era: Fluctuating Seas and Evolution

Cambrian Period

The sun rose the first day of the Cambrian Period, 541 million years ago, on a world geography in disarray. The massive supercontinent of Rodinia was breaking up into individual landmasses. Each terrestrial plate was in constant motion, at a speed comparable to the growth rate of the human fingernail. The slow tectonic dance affected atmospheric and oceanic circulation patterns and created environments conducive to the rapid evolution of plants and animals. The Great Cambrian Explosion was underway, an abrupt appearance of a wide variety of large, multicellular, skeletal life-forms, including representatives of every modern animal phylum with a three-part anatomy—head, body, and tail—that coordinates the functions of sense, energy, and motion.

Ohio was positioned approximately 20 degrees south of the equator and was part of the protocontinent Laurentia, the predecessor of today's North America. Terrestrial vegetation was absent and sunshine relentlessly bathed the naked land. The lack of a water-retaining organic soil allowed the erosive forces of air and water to deeply sculpt the terrain into furrowed profiles. Then, the surface underwent a face-lift as a shallow, warm-water sea slowly spread across the landscape, depositing first sand and mud and later carbonate sediment that was gradually compressed into sandstone, shale, and limestone.

A rich and varied suite of marine life existed at this time, but most phyla were represented by only a few species. The aquatic community consisted principally of brachiopods, corals, sponges, and trilobites. Trilobites made up about half of the total biomass and, by the very nature of their numbers, occupied the top of the food chain. The inch-to-foot-long mobile creatures ingested food from the mud and silt deposits that covered the seafloor, reached their peak of dominance in late Cambrian time, and then declined gradually in numbers. Some scientists suggest a temporary cooling of the seas may have been responsible for this change.

The Cambrian Period is famously known as the time when fauna experimented with the development of mineralized exoskeletons constructed either of calcium carbonate or calcium phosphate. During Precambrian time, saltwater fauna had lived and evolved over millions of years as vulnerable soft-bodied organisms. The formation of an exoskeleton was important because it protected developing marine organisms from a host of dangers: ultraviolet radiation, predators, and death by dehydration in fluctuating, shallow-water environments.

As Cambrian time drew to an end, Laurentia was uplifted, causing a general retreat of the shallow seas from the region of Ohio. Sandstone and dolomite were deposited in the remaining low-lying subtidal areas.

Ordovician Period

Throughout the Ordovician Period, 485 to 444 million years ago, plate tectonic forces were incrementally moving Laurentia in a counterclockwise motion, an action that effectively moved the Buckeye State a bit closer to the equator and edged it into an ever-warmer environment. The majority of Laurentia remained exposed above sea level, so erosion was the dominant force affecting its surface. A complex array of landscapes developed, including caves, hills, and valleys.

By Late Ordovician time, a universal rise in sea level resulted in the formation of the Tippecanoe Sea, an extensive body of saltwater that covered the landmass of Laurentia from Mexico to the Arctic islands. Ohio lay submerged beneath this shallow-water, tropical sea, perhaps the greatest-ever inundation of a continental landmass in the entirety of geologic history.

Off the east coast of Laurentia, subterranean pods of magma migrated to the surface, adding fuel to tectonic forces that were feeding an arc of volcanic islands. Rising to heights of 12,000 feet and stretching from Nova Scotia to Georgia, this volcanic arc was the sole body of land found in the Tippecanoe Sea. The volcanoes deposited great volumes of ash across and within the Tippecanoe Sea. Eventually, tectonic forces accreted this arc to the edge of Laurentia, resulting in the Taconic Mountains, which defined the southeastern border of Laurentia.

Mud was the principal deposit offshore of Laurentia, while soupy and slimy ooze composed of as much as 80 to 90 percent calcareous skeletal remains accumulated in clearer, shallower waters. The offshore mud contained voluminous organic matter that was later heated during deep burial and then converted into subterranean reservoirs of hydrocarbons.

Surface-feeding organisms replaced bottom-feeding types, and a diverse population of bryozoans and corals built relatively small patch reefs that gave character to an otherwise nondescript seafloor. The tepid, oxygenated sea created a utopia for marine organisms—in life they flourished, and in death they were preserved en masse. As a result, the river valleys and roadcuts of southwestern Ohio are world famous for the abundance, variety, and exceptional preservation of the fossils they contain. It has been suggested that if all the fossils were removed from the Ordovician strata beneath Cincinnati, the city would slump below sea level.

More than five hundred families of life thrived in the expansive Tippecanoe Sea. The most abundant were clusters of bryozoans (moss animals), followed by scatterings of solitary brachiopods (lampshells). Pelecypods (clams), cephalopods (squid), and gastropods (snails), all 1 inch or so in size, abounded. Trilobite species declined in number, maintaining a trend that began during Cambrian time.

The end of Ordovician time is marked by the decimation of many groups of organisms, especially bryozoans and brachiopods. This extinction was probably caused by glaciation centered on lands at the South Pole.

Silurian Period

During the Silurian Period, 444 to 419 million years ago, a tranquil environment developed across the length and breadth of Laurentia. The tropical climate was balmy and arid, and the volcanic activity that characterized the rise of the Taconic Mountains was on the decline. These conditions were very conducive to the deposition of limestone in the Tippecanoe Sea, accompanied by an explosive growth of reefs—extensive, layered, ridge-like structures built in warm, clear, shallow water by sedentary calcareous organisms, most commonly colonial corals. Silurian-age rocks in northwestern Ohio are famous for their fossil reefs.

The erosion of the Taconic Mountains created vast deposits of sandstone and shale that were shed across New York, Pennsylvania, and West Virginia and into Ohio. Today these rocks harbor widespread reservoirs of crude oil and natural gas.

Dolomite is the most common type of rock in Ohio from this time. The dolomite was originally deposited in the form of fossil-bearing limestone (calcium carbonate), but migrating volumes of magnesium-bearing seawater and groundwater changed the rock chemistry, made the rock more porous, and destroyed the bulk of the limestone's fossils.

The growth of massive coral reefs created a restricted embayment of the Tippecanoe Sea that extended across the entire northeastern portion of Ohio. When this briny body of water began to evaporate, it formed thick deposits of halite and anhydrite that today extend north under Lake Erie.

Life in the depths of the Tippecanoe Sea remained relatively unchanged from that of Ordovician time, but several new species evolved that menaced anything that moved. Among the new species was a shark characterized by a movable jaw, an extended spine, and body-covering scales. Land-based plants developed for the first time.

Simple vascular plants, characterized by a system of roots and leaves, brought an organic softness to the landscape and opened the door for the arrival of amphibians. The ancestors of plants initially evolved in a marine environment and slowly migrated onto a terrestrial home

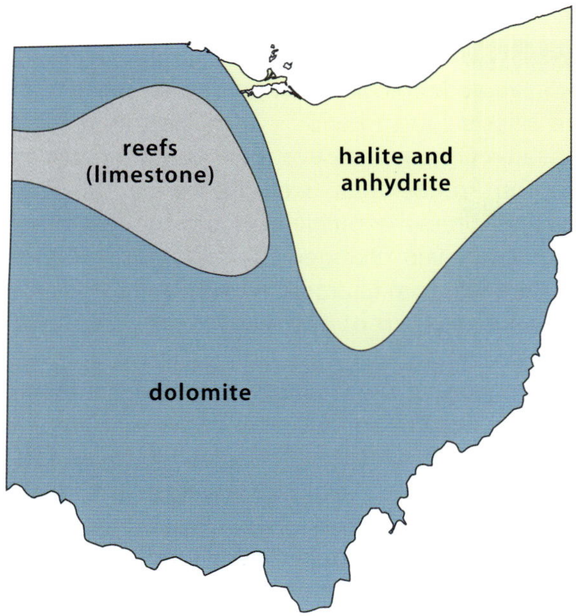

Contrasting areas of deposition in the Tippecanoe Sea 430 million years ago, during mid-Silurian time.

after passing through a freshwater environment. With the development of an organic-based soil, rainfall was more easily absorbed as groundwater, thereby reducing the volume of surface runoff. Erosion was brought under control.

As the sun set on the Silurian Period, the shoreline of the Tippecanoe Sea began to retreat to the south, and the rumblings of tectonic forces heralded a new episode of mountain building to the east.

Devonian Period

The tranquil environment that characterized the Silurian Period ended. Dense sulfur-bearing clouds rolled across Ohio from the east as the dormant magma chambers beneath the eroded roots of the Taconic Mountains were rejuvenated. New generations of volcanoes erupted, and earthquakes fractured the land with increased frequency and intensity. By Middle Devonian time, the growth of the Acadian Mountains was in full stride. These mountains developed as proto-Eurasia, one of the continents that had been created when Rodinia broke up at the end of the Precambrian Era, collided with the east coast of Laurentia. This mountain building event, or orogeny, was the second phase of the three-step tectonic process that built the Appalachian Mountains. The first phase was the Taconic Orogeny.

As the Acadian Mountains rose to greater and greater heights, their limits expanded from Florida north to the Canadian Maritime Provinces. Rivers draining the western Acadian flanks deposited thick sequences of organic-rich silt, mud, and clay across the entirety of Ohio. This influx of sediment filled in the ever-lowering waters of the Tippecanoe Sea, leaving behind extensive deposits of shale and siltstone.

Devonian rocks in Ohio rival the state's Ordovician strata in terms of their fossil abundance and diversity. Brachiopods, clams, snails, and some trilobites are widely distributed. Studies of microscopic growth-ring patterns ornamenting the surface of rugose corals indicate that a 410-day year defined the Devonian calendar, a reduction of 20 days over the prior 100 million years. The rotational velocity of Earth was declining, so a typical Devonian day in Ohio lasted approximately 21.5 hours.

Two groups of bony fish moved into the freshwater realm: the ray-fin, which would become the dominant fish of the contemporary world, and the lobe-fin, which would give rise to the four-legged amphibians that blazed the transition from water to land. The Age of Fishes was in full swim. Although not common, fossil fish remains, scales, teeth, and the occasional armored head and jaw can be found in Devonian strata in Ohio.

Onshore, ground cover shrubs and early trees began to clothe the naked landscape, evidence of the evolving green revolution. Then, some 360 million years ago, surface ocean temperatures plummeted as much as 15°F and 70 percent of the warm-water invertebrate species went extinct. Reef communities were severely impacted.

About 390 million years ago, during the Devonian Period, a headland of deltas and marshes (brown), formed by rivers (dashed lines) draining the Acadian Mountains, gradually filled in the Tippecanoe Sea. Western and northern Ohio were high and dry.

The curtain closed on the Devonian Period with the end of the Acadian Orogeny.

Mississippian Period

The Mississippian Period began 359 million years ago and lasted for 36 million years. Throughout most of the world, Mississippian-age strata are indistinguishable from those of the Pennsylvanian Period that followed; thus the two time periods are often lumped together and called the Carboniferous. In the United States, however, Carboniferous-age rocks are easy to differentiate and are therefore conventionally divided into the Mississippian and Pennsylvanian systems on the basis of, respectively, the relative absence or presence of coal.

The Mississippian Period marks the beginning of the Alleghanian Orogeny, an event of mountain building that rivaled the Precambrian Grenville Orogeny in intensity. During this time the building of the modern-day Appalachian Mountains changed the topographic surface of North America from Labrador to Alabama.

For the better part of 130 million years, ever since the sun had set on the final days of the Cambrian Period, the geologic history of Ohio had been closely associated with an aquatic environment characterized as shallow, marine, tropical, and teeming with varieties of organisms continuously undergoing evolutionary experimentation. On occasion, the geologic effects of climate change, mountain building, and the global migration of protocontinents dealt a near deathblow to the marine community, but life somehow prevailed.

Sedimentation across the eastern margin of Laurentia began to change from predominantly marine to principally nonmarine during the Mississippian. In eastern Ohio, older limestone and shale deposits gave way to thick layers of silt and sand deposited in a series of deltas by rivers flowing off the western flanks of the rising Appalachian Mountains. Simultaneously, in the western part of the state, the waters of a waning epicontinental sea (situated on the shelf or interior of a continent) were increasingly confined to isolated locales as the land rose in response to distant tectonic activity. The limestone deposited in these embayments contains fragmented fossils and ripple marks—evidence that suggests the geographic conditions ranged from very shallow water to tidal flats. Ohio was within 5 degrees of the equator.

The Mississippian marine community reacted to the effects of Devonian extinction with renewed attempts at diversification. Brachiopods quickly recovered, while other groups, such as crinoids and bryozoans, reached a peak of diversity. The great reef complexes that formerly dominated the landscape were replaced by smaller colonies of limited extent.

Ice sheets waxed and waned across the southern hemisphere, but the average global temperature hovered around a moderate 54°F. All across Laurentia, life took advantage of this benign climate, and the biologic spotlight moved from fauna to flora. Vegetation was changing the complexion of Earth.

Pennsylvanian Period

During the Pennsylvanian Period, the Appalachian Mountains reached their full topographic glory as lofty highlands that defined the margin of eastern Laurentia. Turgid rivers flowing to the west transported a huge volume of pure quartz sediment that was compacted into sandstone and conglomerate. Later, erosion reconfigured these rocks into an embankment of scenic cliffs and caves that today cross Ohio from Geauga County south into Pike County.

Early on during the Pennsylvanian Period, shallow seas attempted to again flood the state, but a concert of geologic processes affected their advancement. Fluctuating sea level, related to the growth and melting of glaciers in the southern hemisphere, interacted with episodic mountain building, caused by the ongoing collision of Laurentia with the protocontinents of Europe and Africa, to change the face of eastern Ohio. A complex system of near-shore and coastal environments composed of swamps, lakes, abandoned river channels, marshes, and stagnant lagoons and bays dominated the landscape. The Everglades of Florida and the Okefenokee Swamp of Georgia are modern-day examples of such wetlands.

Plants had been present since the early Silurian Period, when Ohio was positioned 10 degrees south of the equator. Now that plate tectonic motion had moved the state to a near-equatorial location, vegetative conditions had evolved from being primitive and scattered to a state both lush and expansive. The wetlands were blanketed with ferns, reeds, and horsetail rushes, and canopies of conifers and scale trees reached heights in excess of 100 feet. When this vegetation died, it was compressed and heated underground to form Ohio's extensive deposits of bituminous coal.

Repeated sequences of thinly bedded limestone, shale, clay, sandstone, and vegetation (later turned to coal) were deposited in these wetlands. During the Pennsylvanian Period, between 323 and 299 million years ago, more than one hundred individual beds of coal were added to the rock inventory of Ohio. Some sixty of these are extensive and thick enough to have economic significance.

The boggy nature of the lowlands, which could periodically dry out, encouraged the development of both the durable, shelled amniotic egg and coniferous trees, which preferred higher and drier ground. Plant and animal life were thus no longer dependent on an aquatic environment for reproduction, and the door opened for the arrival of amphibians and reptiles. By the end of the period, the seas had retreated from the state, never to return.

Permian Period

The Permian Period, spanning from 299 to 252 million years ago, was a time of historic occurrences and conclusive events that collectively defined the end of the Paleozoic Era. The tectonic crescendo known as the Alleghanian Orogeny raised the Appalachian Mountains to elevations that may have exceeded those of the present-day Himalayan range, while simultaneously the greatest mass extinction event ever recorded affected Earth's life-forms.

For the first time since Rodinia had coalesced more than 0.7 billion years earlier, the assembly of a new supercontinent was nearly complete. Shaped like the letter C, Pangaea—"one Earth"—extended from pole to pole and was distinguished by two youthful mountain ranges: the Urals of Asia and the Appalachians of North America.

The Appalachian Mountains delineate the suture along which proto-Europe and proto-Africa reached the final state of collision with Laurentia. The geologic forces that built this range also elevated lands to the west, beyond Ohio as far as the longitude of Colorado. Sedimentary rock was still being deposited in the Buckeye State, but it was entirely nonmarine in composition: sandstone and shale in river channels, mudstone and limestone in freshwater lakes, and a few scattered coal beds. Today, these rocks are exposed in the half-dozen counties that define the extreme southeastern sector of the state.

Permian-age rocks of Ohio are not very fossiliferous: reptile and amphibian bones, some freshwater invertebrates, and plant and insect parts have been discovered, but the record is sparse and very incomplete. This scarcity of fossils is directly related to an event that brought the Permian Period and the Paleozoic Era to an end: 95 percent of all marine invertebrate species and 70 percent of their terrestrial brethren became extinct. The cause of this biologic calamity is still being debated. Did Earth collide with an extraterrestrial body, causing worldwide ecological change, or was the culprit widespread episodes of volcanism that released immense clouds of ash that, in turn, sullied the clear-water environments of the oceans? Regardless of which theory is correct, the all-encompassing assembly of Earth's landmasses at the end of the Permian Period was nearly fatal to life on Earth.

An Interim Period

From the end of the Permian Period to the present, Ohio has been positioned well above sea level. The primary geologic process in operation has been erosion, caused by the forces of wind, running water, gravity, and moving ice. Strata representing the entirety of the Mesozoic Era and much of the Cenozoic Era are thus nowhere to be found in the state. This missing rock record, extending from 252 million years ago to the initiation of the ice ages 2.6 million years ago, is a void in the geologic history of Ohio that can only be surmised, never written.

Throughout this extensive period of time, Ohio continued to be on the geographic move, slowly and continually driven north by plate tectonic forces. By the dawning of the Pleistocene Epoch, it had reached a position of 40 degrees north latitude, a location some 60 degrees away from its starting point way back in Precambrian time. It had been a long trip, but the Buckeye State had finally reached its present-day geographic location, ready to experience yet one more onslaught of geologic activity, one that would extensively scar its surface.

III. Pleistocene Epoch: Ice and More Ice

The shortest chapter in the geologic history of Ohio began 2.6 million years ago when, in direct response to the cooling of Earth's climate, snow and ice accumulated in the Hudson Bay area to a depth sufficient to form a massive ice sheet that spread outward in all directions under the force of its own weight. Two-thirds of Ohio was eventually enveloped in the chilly embrace of ice estimated to have grown more than 1 mile thick.

Historically, the ice age of North America has been discussed in terms of four major glacial advances; from oldest to youngest they are the Nebraskan, Kansan, Illinoian, and Wisconsinan. Today, geologists recognize some thirty overall cycles of ice advancement and retreat of one degree or another, and the two major episodes that occurred before the Illinoian are combined and referred to as the pre-Illinoian. Thus the ice age was actually composed of multiple ice *ages*. The Wisconsinan stage, the best studied of the four, began 110,000 years ago, entered Ohio 24,000 years ago, and had left the state 10,000 years later.

U-shaped valleys gouged out by ice, the five Great Lakes, outwash plains, moraines, and abundant landscapes of fertile soil identify the Pleistocene footprint in North America. As the final ice sheet retreated across Ohio, great volumes of meltwater deposited layers of drift valued today as sources of sand, gravel, and potable groundwater. Sediment deposited in tunnels beneath the ice was left behind as eskers, and beach ridges formed

Glaciers repeatedly bulldozed two-thirds of Ohio (blue area) during the Pleistocene Epoch.

along each of the several fluctuating shorelines of the lake that eventually stabilized as Lake Erie. The most topographically prominent of the glacial features are the many moraines that accentuate the northeastern portion of the state. These arcuate ridges mark positions where the retreating ice temporarily stagnated, and the melting ice released masses of sediment it had collected during its lengthy journey from the north.

Many of the animals and plants that called Ohio home during the Pleistocene Epoch are similar to those that inhabit Ohio today. However, the change in vegetation that occurred when the climate warmed some 10,000 years ago is thought to be primarily responsible for the extinction of many ice age animals. The evidence of the former presence of these animals—mastodons, woolly mammoths, short-faced bears, giant beavers, giant ground sloths, and long-horned bison—is found in the many fossil bones that have been found throughout the state.

Epilogue

Over the past two centuries, geologists have determined the geologic history of Ohio by studying rock associated with the Precambrian Era (involving some 700 million years), the Paleozoic Era (289 million years), and the Pleistocene Epoch (less than 3 million years). The sum of these three time segments is approximately 992 million years, a mere 22 percent of the 4.54 billion years that represent the age of the Earth. The record of the geologic history of Ohio is, therefore, like a 100-page book that contains 22 pages of text intermixed with 78 blank pages.

Nevertheless, much history has been determined. Ohio was born during the time when ancient minicontinents crystallized and then merged into larger continents; it progressed through youthfulness, when life evolved from ocean to land and the Appalachian Mountains gained stature, and it matured when ice sheets scoured its landscape.

The study of geology is largely dictated by the concept that "the present is the key to the past." Turning this statement inside out—"the past is the key to the future"—gives credence to the prediction that Ohio will continue to be affected into the distant future by the ever-present geologic forces of construction and destruction.

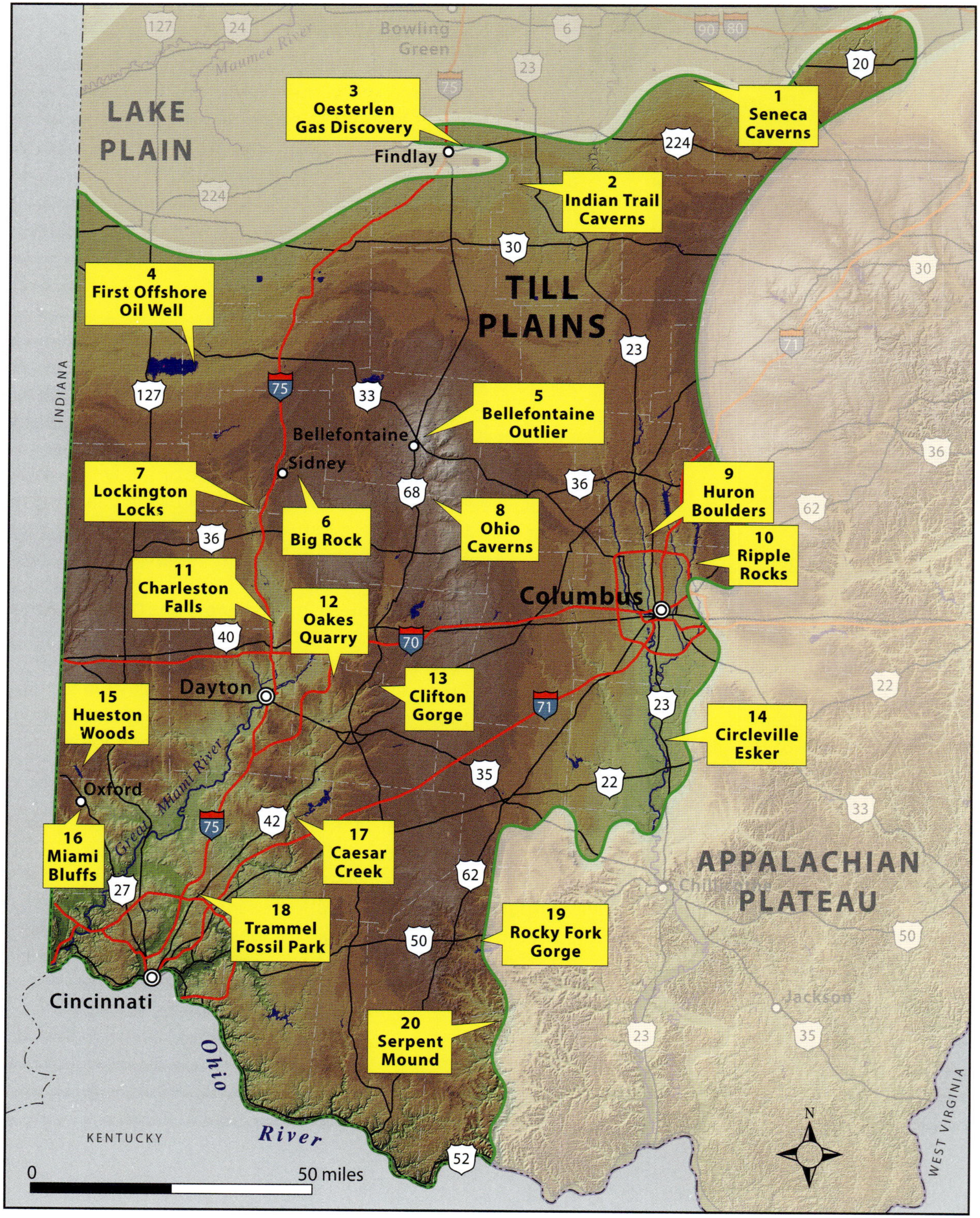

LAKE PLAIN
TILL PLAINS
APPALACHIAN PLATEAU
Bowling Green
Maumee River
Findlay
Bellefontaine
Sidney
Columbus
Dayton
Oxford
Cincinnati
Chillicothe
Jackson
Great Miami River
Ohio River
INDIANA
KENTUCKY
WEST VIRGINIA
1 Seneca Caverns
2 Indian Trail Caverns
3 Oesterlen Gas Discovery
4 First Offshore Oil Well
5 Bellefontaine Outlier
6 Big Rock
7 Lockington Locks
8 Ohio Caverns
9 Huron Boulders
10 Ripple Rocks
11 Charleston Falls
12 Oakes Quarry
13 Clifton Gorge
14 Circleville Esker
15 Hueston Woods
16 Miami Bluffs
17 Caesar Creek
18 Trammel Fossil Park
19 Rocky Fork Gorge
20 Serpent Mound
0
50 miles
N

TILL PLAINS

The Till Plains physiographic province of western Ohio consists of rolling landscapes accentuated by clusters of curvilinear moraines—mounds of rock and soil, up to 130 feet high and 6 to 10 miles long, deposited by melting ice at the edge of a glacier. The moraines overlie thick blankets of fertile till that harbor significant resources of potable water and support an economically important agricultural industry. These unstratified and unconsolidated mixtures of clay, sand, and gravel were deposited by glacial meltwater.

During presettlement times, the mineral-rich glacial soils nourished tall prairie grass and extensive regions of hardwood forests. The highest point in the Till Plains province is Campbell Hill, an oval-shaped ridge that rises 1,549 feet above sea level in the city of Bellefontaine.

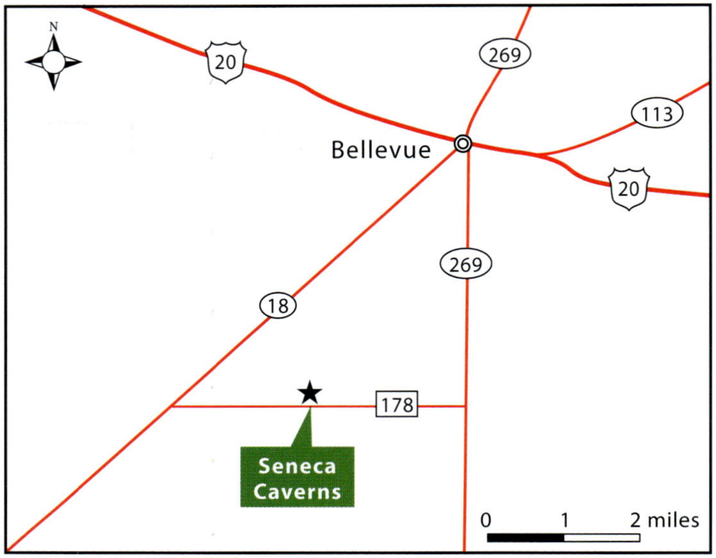

1. SENECA CAVERNS
◆ *Caviest Cave in the USA*

Not all caves are created equal. The solution variety, formed when acidic groundwater dissolves carbonate rock, is the most common; some five hundred are known in the Buckeye State. Seneca Caverns, located 4 miles southwest of the community of Bellevue, belong to a different category. Geologists have mapped seven levels of contrasting dimensions, all connected with a fracture that descends at a 40-degree angle. Five additional levels are exposed during periods of drought, when groundwater levels sink to greater than normal depths. The angular and unweathered nature of the bedrock coupled with the absence of massive cave-lining stalagmites and stalactites are evidence that these caves were not created by percolating, acidic groundwater. What then, is the origin of Seneca Caverns?

Two sets of fractures—one trending northeast, the other northwest—break the sedimentary bedrock of Seneca County. Seneca Caverns are located in limestone and dolostone along one of the larger northwest-oriented fractures, suggesting a cause-and-effect relationship. Studies have shown that the conversion of deep-seated Silurian-age anhydrite to gypsum deformed the overlying Devonian-age rocks to the degree they were extensively fractured. Groundwater moving along the fractures dissolved the gypsum, causing the Devonian-age rocks to collapse into the voids. Seneca Caverns are thus classed as collapse caverns.

Angular blocks of limestone and dolostone litter the cavern floor and appear to fit into ceiling recesses, much like a three-dimensional puzzle. The shattered subterranean maze descends to a known twelfth level, but at the seventh, Ole Mist'ry River normally ripples along the floor, flowing in a northerly direction to discharge into Lake Erie. Advertised as the Earth Crack, this electrically lighted cavern system is the only one of its type open to the American public.

Passageways littered with limestone and dolostone blocks are evidence of the collapsed nature of this multilevel crack in the Earth. —Courtesy of Gregory J. Kemp

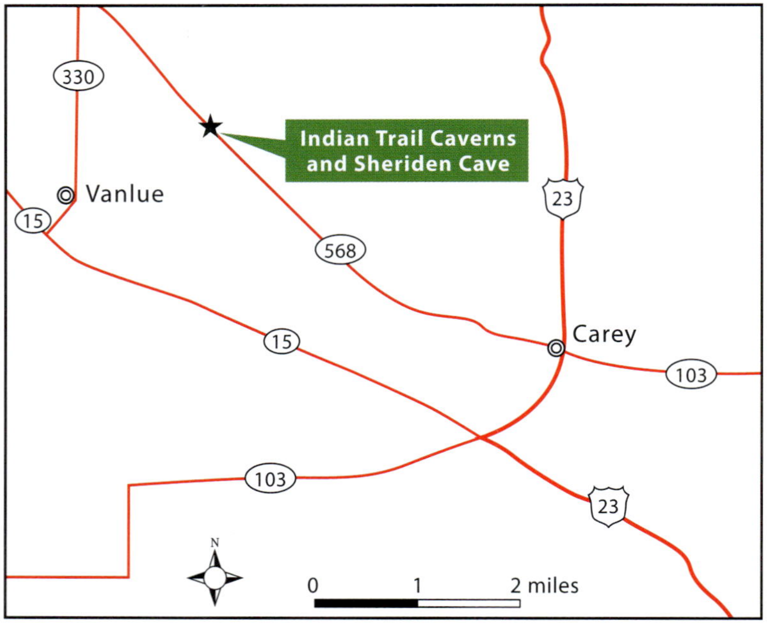

2. INDIAN TRAIL CAVERNS
Ice Age Death Trap

The undulating tableland of western Ohio seems an unlikely location for a chain of bedrock voids that, over a period of centuries, spelled death to a remarkable variety of large and small animals, many of which are now extinct. One does exist, however, trending north to south across a ridge of Silurian-age dolomite that accentuates the topographic profile of northwestern Wyandot County. This chambered chasm of death consists of three associated caverns: Sheriden Cave, Hendricks Cave, and the nearly 600-foot-long Indian Trail Caverns. Subterranean investigations suggest all belong to

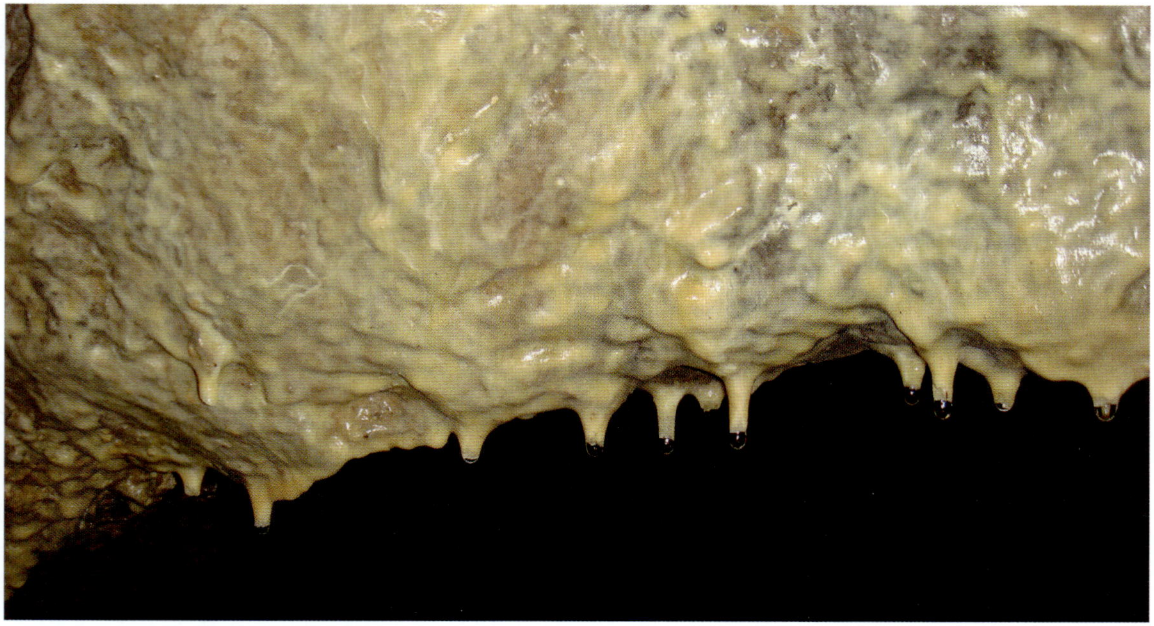

Delicate soda-straw speleothems cling to the ceiling at Indian Trail Caverns. —Courtesy of Keith Hendricks, Indian Trail Caverns

one cavern system, which formed through the dissolution of dolomite by acidic groundwater flowing through bedrock fractures.

Later, after the last of the Pleistocene ice sheets had retreated from Ohio, unconsolidated sediment periodically filled portions of the cavern system and buried the remains of a variety of inquisitive mammals that had either fallen into or sought shelter within its chambers. The disarticulated bones of at least ten humans, ages fifteen to fifty, have been found in Hendricks Cave. In contrast, the fossil remains of at least sixty different animals have been discovered in nearby Sheriden Cave. This treasure trove includes stag-moose, short-faced bear, giant beaver, and flat-headed peccary—all now extinct.

The sediment fill of Indian Trail Caverns was removed so this extended chamber could be opened to public tours on a seasonal basis. Air cascading through natural portals creates a constant temperature of 55°F, and skylights illuminate young, delicate soda-straw speleothems hanging from the ceiling. Measurements taken over the past half century show that these pygmy stalactites are growing at the rate of 0.1 inch per decade.

Small in presence, but big in paleontological value, these caves function as tiny windows that permit a magnificent modern-day view into the often mysterious annals of a vanished ice age.

Light streaming through a skylight at Indian Trail Caverns.
—Courtesy of Keith Hendricks, Indian Trail Caverns

3. OESTERLEN GAS DISCOVERY
Northwest Ohio Gas Boom

Dr. Charles Oesterlen was a visionary and an optimist, and in 1884 he drilled a well 1 mile east of downtown Findlay. When the borehole reached 1,092 feet, he decided to "shoot" it with 30 quarts of nitroglycerine. The explosion produced a flow of natural gas that, when ignited, was seen 15 miles away. Soon the well was producing 250,000 cubic feet of gas every 24 hours—enough to warm the average American home of 2013 for 2.5 years. The resulting drilling boom engulfed northwestern Ohio and was soon being touted in headlines throughout the nation. Oesterlen, a licensed physician and amateur geologist, had been right all along: there was wealth in the Ordovician-age Trenton Limestone that formed the rock basement of Hancock County.

Awakened to the potential, civic leaders hired a public relations guru, and soon countrywide advertisements promised free gas to any manufacturer that would relocate to Findlay. When the management of a dozen glass factories responded, the population exploded from 4,500 to 25,000 individuals. By 1889, however, production was on a serious decline, and the gas boom was all but history.

Speculators believed an inexhaustible supply of natural gas was entombed in the Trenton Limestone, and uncontrolled production had become the extravagance of the day. Gas from producing wells flowed on a 24/7 basis, and flares burned everywhere in Findlay—in homes, businesses, and factories. To the delight of community merchants, Main Street was continually illuminated by the glare of fifty-eight gaslit arches.

A buried storage tank and adjacent borehole on the grounds of the Hancock County Fairgrounds identify the site of the Oesterlen discovery well.

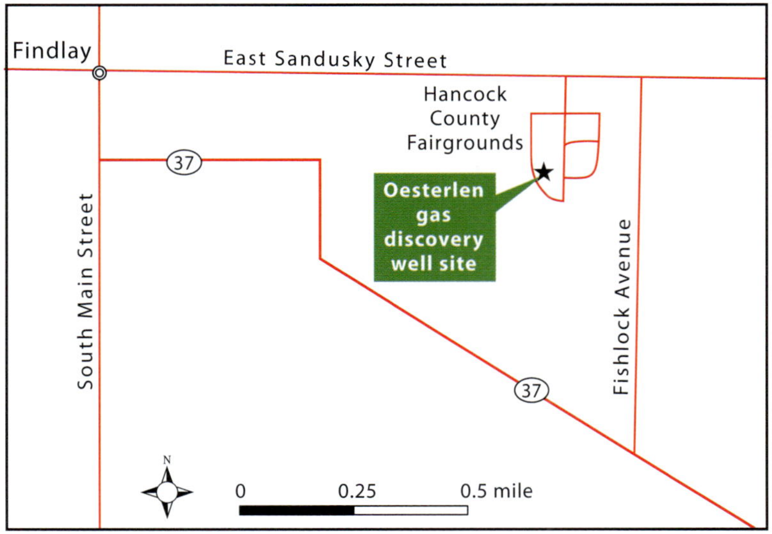

The Oesterlen discovery was the first of an estimated 76,000 wells that eventually produced 380 million barrels of oil and an estimated 1 trillion cubic feet of gas in Ohio between 1885 and 1935.

4. FIRST OFFSHORE OIL WELL
A Field of Contenders

When any assembly of individuals involved in the exploration of crude oil begins discussing the drilling of the first offshore well, a spirited debate commonly ensues. All agree the event took place within the forty-eight contiguous American states, but there is not agreement concerning which of the four contenders should have the title.

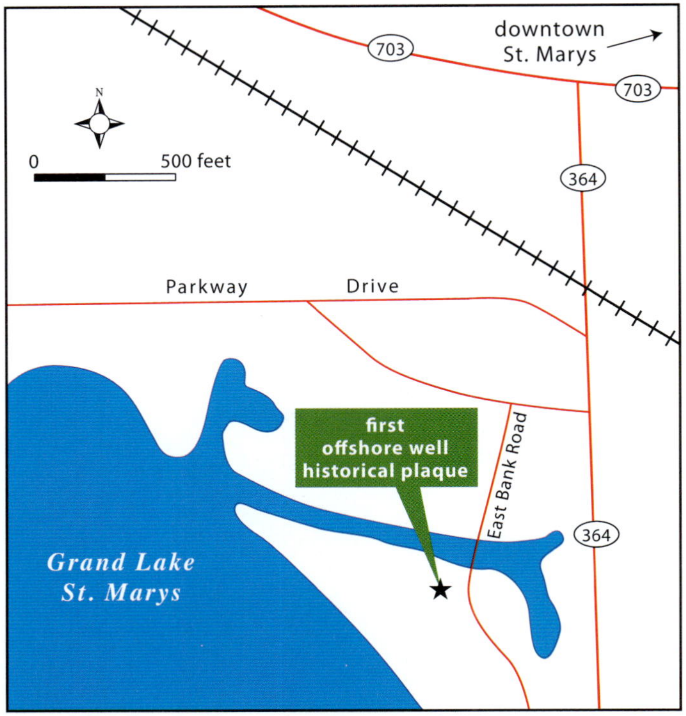

Pen-and-ink interpretation of the world's first offshore well.
—Drawing by Patsy Faires, from a copy of a historic photograph courtesy of George Neargarder

This historic plaque documents that the very first offshore well was drilled in Grand Lake St. Marys, immediately beyond the central pair of trees.

The Superior-Pure State No. 1 oil-producing borehole, drilled 1 mile offshore in the Gulf of Mexico in 1937, is often touted as the winner. This claim doesn't sit well with residents of northwestern Louisiana, who brag that Gulf Oil drilled the Ferry Lake No. 1 in the waters of Caddo Lake in 1911, the first of hundreds of oil wells that eventually dotted the lake. Yet another claim is associated with the 1896 discovery of the Summerland Oil Field in the waters of the Pacific Ocean, offshore of California.

In fact, a thorough search of historic records indicates that each of these operations must settle for wannabe status. The title of first offshore well properly belongs to a borehole drilled in Grand Lake St. Marys in January 1891. Documented by the Ohio Geological Survey, this well extended the limits of the Lima Field, the first giant oil field to be discovered in North America, many miles to the southwest. A historical plaque on the grounds of Grand Lake St. Marys State Park identifies the precise location of this pioneering project.

To date, some 7,200 oil and gas fields have been discovered in the offshore waters of 127 countries. This global industry was born in the shallow waters of Grand Lake St. Marys, almost a decade before the birth of the twentieth century—a truly decisive case of being first.

5. BELLEFONTAINE OUTLIER
A Rocky Ramification

Generally speaking, Ohio possesses neither altitude nor relief worthy of boastful discussion. The view from atop Campbell Hill, located on the Hi-Point Career Center campus, 0.5 mile east of the intersection of US 33 and Ohio 540, east of Bellefontaine, however, is worth the visit. Billed as the loftiest in the state, this 1,549-foot-high prominence of glacial till caps a landscape of Devonian-age rock entirely surrounded by older Silurian rock.

Some 430 million years ago, a warm, shallow sea inundated Ohio and deposited thick sequences of limestone and shale. Gradually, the ocean environment changed to swamp and then dry land, and erosion replaced deposition

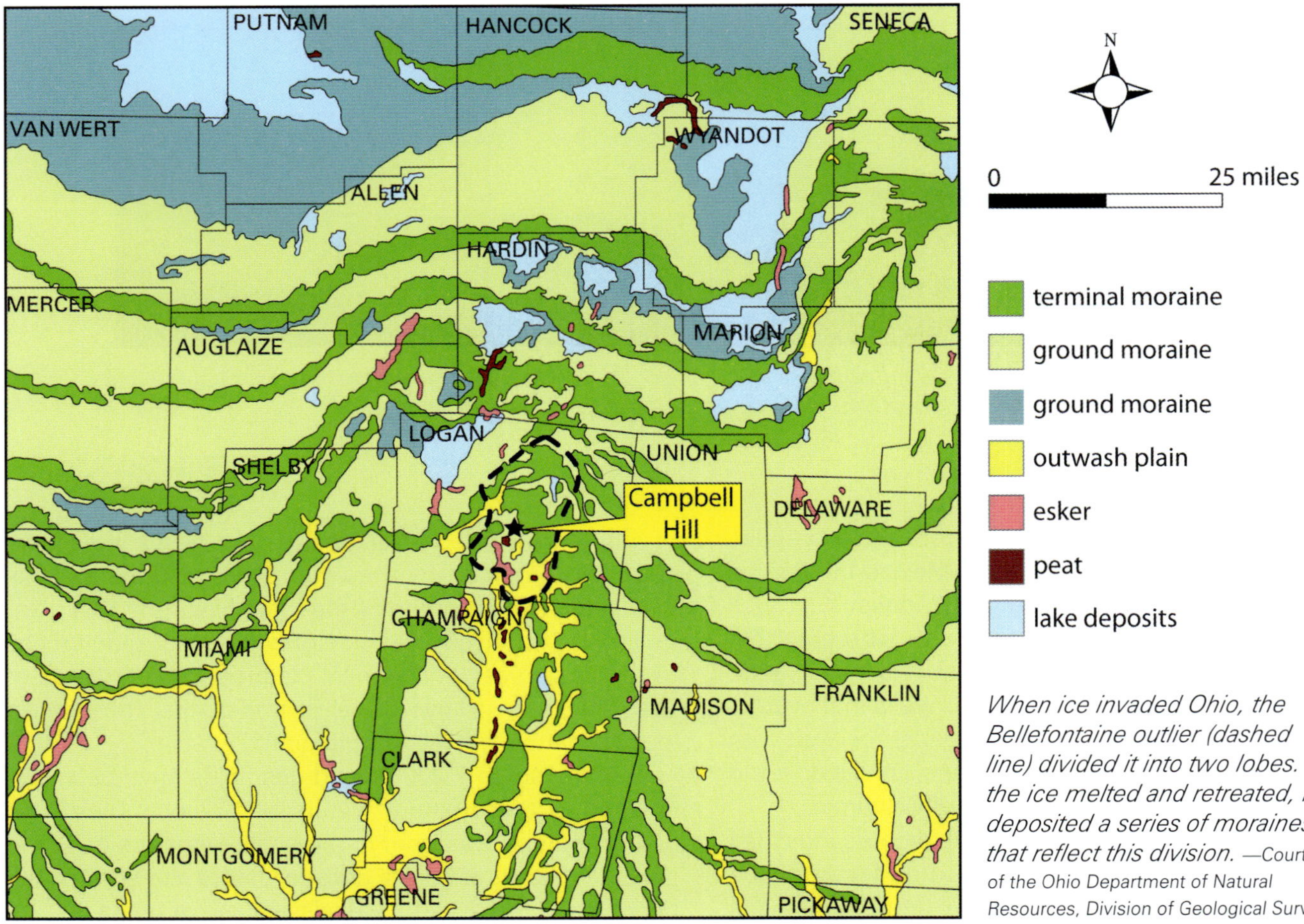

When ice invaded Ohio, the Bellefontaine outlier (dashed line) divided it into two lobes. As the ice melted and retreated, it deposited a series of moraines that reflect this division. —Courtesy of the Ohio Department of Natural Resources, Division of Geological Survey

as the dominant geologic effect. Eventually, erosion removed most, but not all, of the Devonian–age rock from western Ohio. The remnant exposure left covering the majority of Logan County is known as the Bellefontaine outlier. An outlier is an area of isolated rock surrounded by older rock.

Some 2.6 million years ago a 1-mile-thick sheet of ice moved southward from the Hudson Bay region and seemingly destroyed everything in its path. A kind of sole survivor, the Bellefontaine outlier constituted a rock-cored topographic barrier, lying rigid like a geologic traffic cone and forcing the advancing ice sheet to split into two tongues: the Miami glacial lobe to the west and the Scioto glacial lobe to the east.

Consequently, the topographic profile of west-central Ohio is today characterized by a series of curvilinear terminal moraines that cascade from the apex of the Bellefontaine outlier and drape across the countryside like the folds of a curtain. This unusual landscape is the net effect of incomplete erosion, interrupted glacial advance, and the resistance of a mass of rock that remains secure in its isolation.

Campbell Hill, the highest point in Ohio, is adorned with several buildings and a silo.

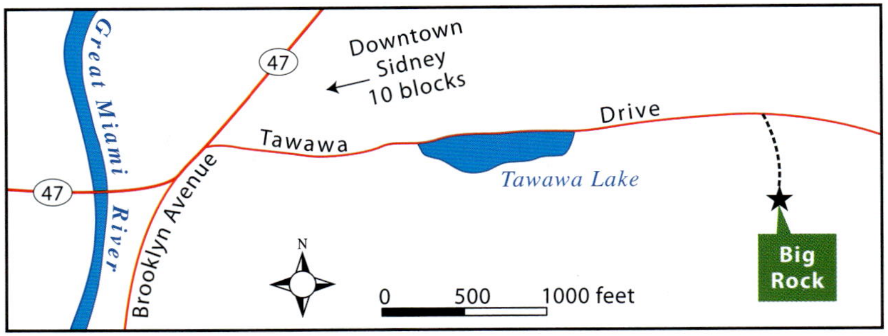

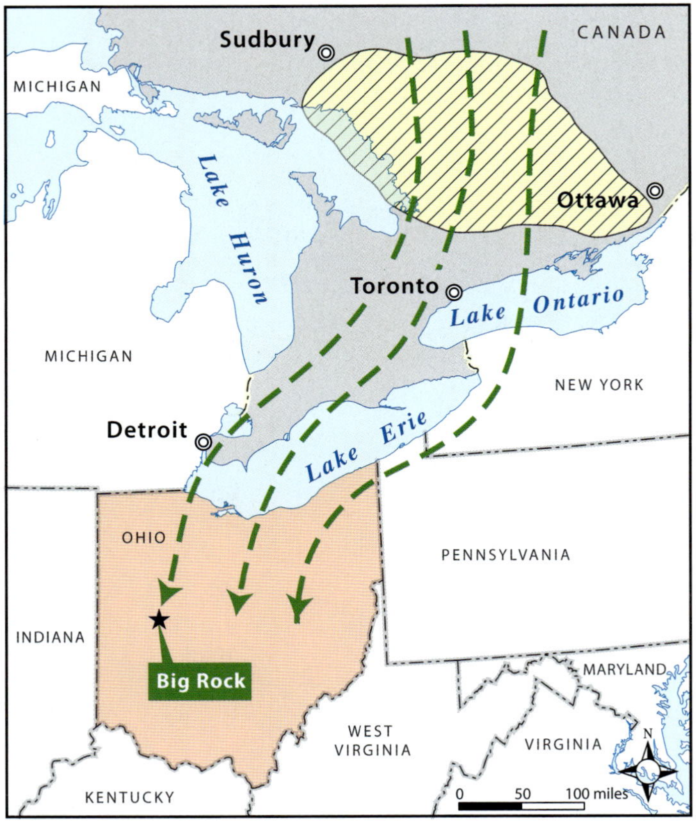

Big Rock was plucked from the bedrock of southeastern Canada by glacial ice, transported south and west into Ohio, and finally laid to rest in Shelby County when the ice melted.

6. BIG ROCK
An Erratic Traveler

Vagabonds of the Earth, isolated boulders of nonnative rock—termed erratics—are haphazardly distributed in soils around the world and come in all sizes, but those suffering from "rock obesity" receive most of the publicity. Called wanderers, orphans, aliens, and immigrants, their presence has been attributed to icebergs, floods, and extraterrestrial arrival. In short, they are *erratic*, but not without explanation.

Consider the rounded, brutish, 1,250-cubic-foot, 103-ton specimen lying sheltered in Tawawa Park, east of Sidney. Named Big Rock, this garden variety of granite is enmeshed in 23,000-year-old ice age soil that overlies limestone bedrock deposited 420 million years ago, during the Silurian Period.

Like all erratics, Big Rock is a rock fragment carried by ice and deposited some distance from the outcrops from which it is derived. It was born when glacial meltwater beneath a glacier penetrated bedrock fractures, froze, expanded, and broke off fragments of rock that were incorporated into the frozen mass. When the glacier reached its limit of advancement and began to melt, the quarried remnants it carried were left behind, testimony to the awesome carrying power of moving ice.

The mineral composition of Big Rock has no counterpart in the bedrock found in Ohio. This is a key characteristic that distinguishes an erratic from any other large rock.

Because granite exposures do not occur in Ohio, Big Rock's parent must lie elsewhere. Studies show that the Pleistocene-age ice sheet that invaded the state on several occasions originated in southeastern Canada and advanced south and west, away from a massive outcrop of 985-million-year-old granite that lies exposed between Ottawa and Sudbury, Ontario. Since its mineral makeup is strikingly comparable to that of Big Rock, this must be the long-sought-after parent.

Quarried from Canada, transported 700 miles southwest, and unceremoniously deposited in Tawawa Park, Big Rock remains today demonstrative evidence of the 1-mile-thick ice sheet that once enveloped the state in its chilling grip.

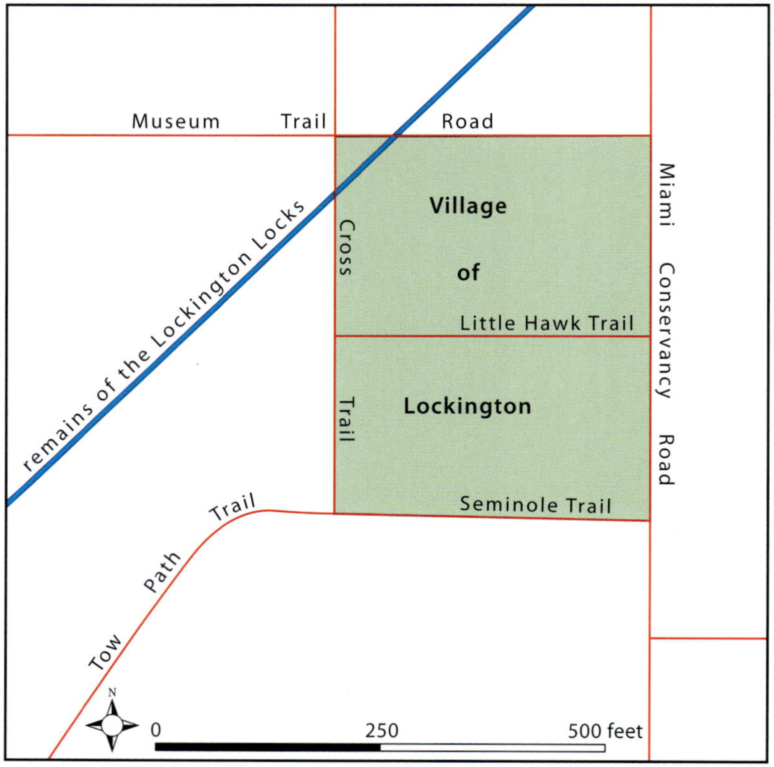

7. LOCKINGTON LOCKS
◆ *A Carbonate Waterway*

The Erie Canal, the transportation system that linked New York City and the Hudson River with Lake Erie, opened in 1825. Soon after, this pioneering waterway-to-the-west was labeled both a financial bonanza and perhaps the most profitable venture in American history. Because success breeds competition, numerous similar projects quickly sprang into operation, especially within the flat terrain found west of the Appalachian Mountains. One such grand scheme was the Miami and Erie Canal, which linked the Great Lakes port city of Toledo with the Ohio River city of Cincinnati. Geology, in the form of topography and bedrock type, played a significant role in the design and construction of this waterway.

Western Ohio is drained by two principle river systems: the Great Miami River, flowing to the south, and the Auglaize and Maumee River system, flowing to the north. The canal was engineered to take advantage of the water in these two drainages. Loramie Summit, a 944-foot-high piece of ground centered in the bucolic village of Lockington, in Shelby County, separates these two drainages. The presence of this drainage divide rooted in carbonate bedrock necessitated the construction of six locks that lifted canal boats a height of 67 feet over a distance of 0.5 mile.

The long-term success and stability of this stair-step system of locks required an erosion-resistant construction rock—and due to economic limitations, one that was accessible. Nearby exposures of the Silurian-age Dayton Formation fit the bill.

The fine-grained nature of this 10-foot-thick, bluish-gray, crystalline dolomite and limestone makes it an excellent building stone. Massive blocks were mined from nearby quarries and hauled to the Lockington construction site by horse-drawn wagons, where they were finished by stonemasons to the degree that they could be fitted together with minimal use of mortar.

Today, a full century after operations along the Miami and Erie Canal were terminated, the Lockington Locks, an engineering facility built of and anchored in carbonate rock, remain testimony to the long-ago era when canals were the transportation system of choice in an expanding America.

Stonemasons refined the blocks of Dayton Formation to the degree that a knife blade cannot be fitted between them.

Lock No. 5 as it appeared after a century of disrepair and before reconstruction had been completed in late 2014.

An intergrown array of translucent stalagmites and stalactites highlight the passageway through the Palace of the Gods. —Courtesy of Tim Grissom, Ohio Caverns

8. OHIO CAVERNS
A Garden of Crystals

Year after year, farmers harvest corn and soybeans from the rich glacial soils of northern Champaign County, providing a treasure vital to the health and sustenance of humanity. Many visitors to this bucolic, rolling landscape are surprised to learn that a third treasured crop is present, one that began to grow tens of thousands of years ago in an environment that today remains a steady 54°F.

Located under Ohio 245, 4 miles east of West Liberty, Ohio Caverns harbors a subterranean forest of elongated crystals that grow along 2 miles of surveyed passageways. Known not only for their size—Crystal King is just shy of being 5 feet long—the hundreds of stalagmites and stalactites exhibit a rainbow range of color that varies from yellow, orange, blue, brown, and reddish-black to pure milk-white.

The geologic history of this natural fairyland is written in three chapters. The first involves the deposition of the Columbus Limestone in a shallow sea that covered Ohio

The aptly named 400-pound Crystal King, the largest stalactite in the Ohio Caverns. —Courtesy of Tim Grissom, Ohio Caverns

390 million years ago, during the Devonian Period. Over the course of millions of years, or chapter two, the land rose and rainwater and groundwater—made acidic through the absorption of carbon dioxide from the atmosphere—slowly dissolved portions of the limestone, forming the caverns. The third chapter began when the groundwater level dropped, leaving the caverns empty. Percolating water began depositing calcium carbonate in the caverns, forming stalactites that slowly grow from the ceiling down. Droplets falling to the cave floor form stalagmites, which grow from the floor up. Studies indicate it takes 500 to 1,000 years for 1 cubic inch of calcite to be added to these crystalline accouterments. The most senior of the crystals is estimated to be 250,000 years old.

Formed over the span of incalculable years, Ohio Caverns is an inviting and photogenic array of iridescent crystals created the old fashioned way: drop by drop, employing a combination of minerals, water, and time—lots and lots of time.

While Ohio Shale concretions up to 9 feet in diameter have been found, this 6-inch-diameter beauty is more typical. —Courtesy of Kendall Hauer, Miami University

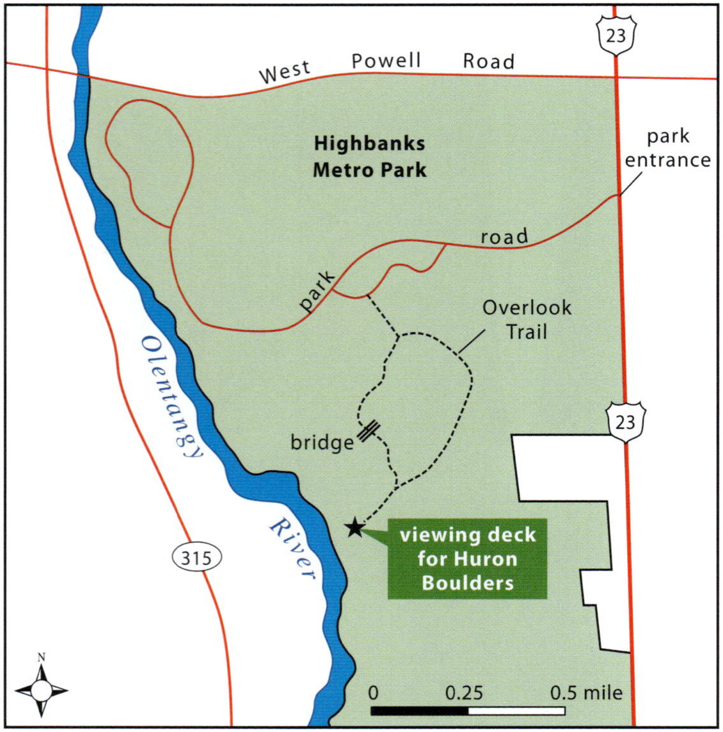

9. HURON BOULDERS
◆ *Puzzling Spheres of Stone*

Perhaps no other rock exposure in the Buckeye State generates as much curiosity as the 100-foot-high bluffs of black Devonian-age Ohio Shale that tower over the Olentangy River in Highbanks Metro Park. The bluffs between the viewing deck and bridge along the Overlook Trail appear to have been the target of a Civil War bombardment. Embedded within and lying at their base are rounded and broken masses of rock that observers often describe as glacial erratics, or even extraterrestrial debris. In actuality, these oversized cannonball-shaped spheres are outstanding examples of uncommon geologic features known as concretions.

Concretions are hard, compact, spherical to irregular masses of sedimentary rock formed by the precipitation of mineral matter around a nucleus that is commonly a fragment of fossil wood, shell, or bone. Generally they are composed of material different from that of the enclosing rock and often exhibit horizontal ribbing that appears to be an extension of the bedding planes of the host rock. The Ohio Shale concretions are composed primarily of calcium carbonate.

While most geologists agree these concretions formed during the interval of time that occurred after the shale was deposited and before it had been completely compacted, the specific chemical process is subject to continuing speculation. One school of thought suggests that a low-density, calcium-rich, waxy slurry composed of the decayed remains of marine life—a natural substance known as adipocere—is the mother substance of concretions.

Throughout central Ohio, Huron Boulders—the name given exclusively to concretions found within the Huron Shale Member of the Ohio Shale because they are generally larger and more perfectly developed than concretions found in the other two members of the formation—are a familiar sight as yard and garden ornaments. Their presence is a reminder that many geologic processes are not yet fully understood.

The layers of Ohio Shale surrounding this Huron Boulder bend around the upper and lower surface of the boulder, suggesting the concretion formed prior to compaction of the shale —Courtesy of Mark Steinmetz

10. RIPPLE ROCKS
A Fossil Seafloor

Once upon a time, hundreds of millions of years ago, the lands that would become Ohio were located 10 degrees south of the equator. Limestone was being deposited in a warm, shallow sea basked by balmy breezes. Coral reefs abounded, sharks were making their appearance on the evolutionary stage, and vegetation was beginning to clothe the landscape. Conditions were good and becoming better for many life-forms. Then, the geography began to change.

To the east, the fender-bender tectonic collision between proto–North America and proto-Europe crafted the Acadian Mountains, the second of three continental impacts that would eventually define the Appalachian Mountains. Rivers flowing to the west formed a vast delta composed of thick layers of sedimentary rock, one of which became the Berea Sandstone.

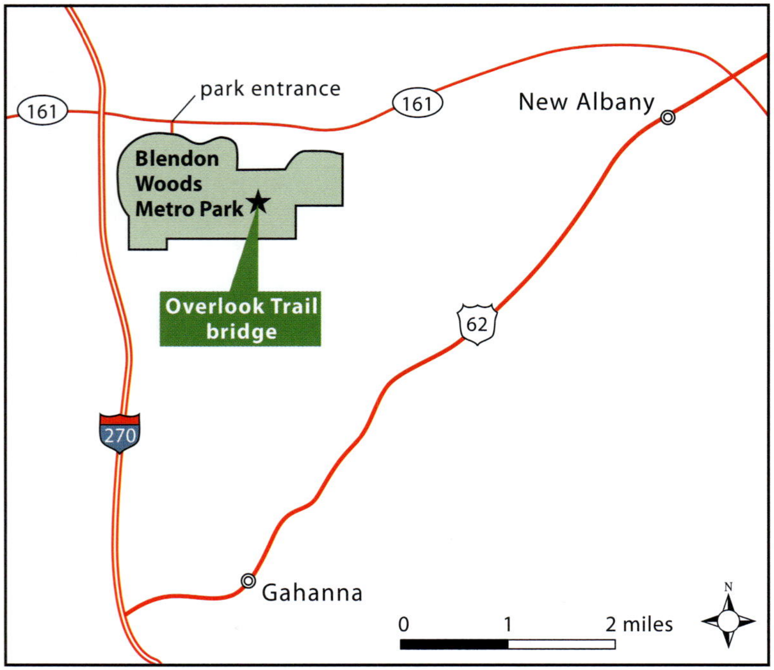

Exquisitely formed ripple marks, indicators of deposition in shallow water, commonly mark Devonian-age Berea Sandstone exposures in Blendon Woods Metro Park, west of New Albany, a northeast suburb of Columbus. Platter-sized fragments of rippled Berea Sandstone accent the bed of Ripple Rock Creek, and a variety of examples can be found under and around the wooden bridge along the 0.6-mile-long Overlook Trail. These ripples testify to an aquatic environment in which the to-and-fro motion of currents shuffled sand-sized particles into a series of evenly spaced ridges.

From a vantage point on the bridge, with a fragment of Berea Sandstone in clear view, one can imagine, with scientific justification, standing in warm, waist-deep, shark-infested waters washing over a coastal plain similar to the Mississippi River Delta of Louisiana. Let there be little doubt about it: you are experiencing an exercise in time travel, standing above a fossil seafloor that formed 370 million years ago. Beware of the high tide. It's due any minute now—and look out for those sharks!

Ripple marks can be produced on land by wind and offshore by waves and currents. The presence of shark teeth in the Berea Sandstone is proof that offshore currents formed the Blendon Woods ripples. Pen for scale. —Courtesy of Helene Blowers

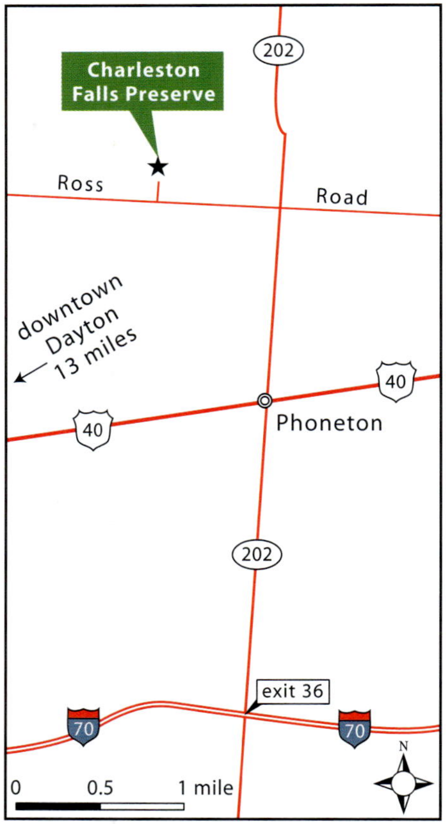

11. CHARLESTON FALLS
The Birth and Death of Waterfalls

Waterfalls are landscapes of wonder composed of plunging assemblages of foam and water. But look quickly, for geologically they all are ephemeral. As a result of recession, the process of upstream retreat, they are constantly changing in profile and position. Charleston Falls, northeast of Dayton, in a nature preserve of the same name, is a case in point.

Formed 16,000 years ago as the last glacier retreated from central Ohio, Charleston Falls owes its 37-foot height and upstream migration to resistant Silurian-age limestone overlying less-resistant Ordovician-age shale. As the pounding force of falling water erodes the shale, overlying layers of limestone, having lost their support, collapse, and the waterfall takes a step in its retreat upstream.

The rock column that forms Charleston Falls, very visible and touchable during the dry season, encapsulates a time when brachiopods and bryozoans ruled the oceans, coral reefs first appeared in abundance, plants invaded the land, and the not overly deep waters of the

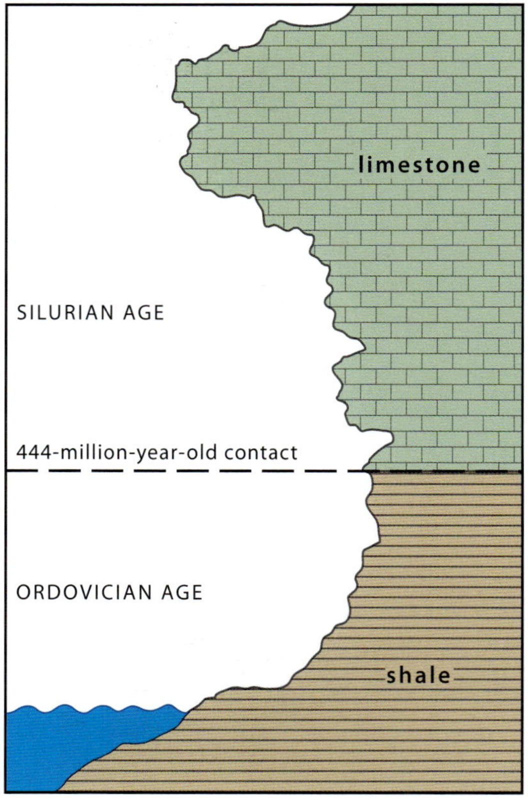

The Ordovician-Silurian contact at Charleston Falls is located at the innermost recess of the waterfall.

Charleston Falls, the "miniature Niagara" of Ohio. The continuing, retreat of the falls is evidenced by the rock rubble seen at the base.
—Courtesy of Jim Telljohann

Tippecanoe Sea washed over the landscape of proto–North America.

Because this shallow sea deposited equivalent amounts of limestone and shale from Nevada to New York, the recession rate of Charleston Falls can be effectively compared to that of American Falls, the New York State side of Niagara Falls. This honeymoon haven of cascading cataracts has retreated 7 miles since its birth 12,000 years ago, evidence of an average recession rate of 3.1 feet per year. In contrast, Charleston Falls has retreated some 5,000 feet from its junction with the Great Miami River in 16,000 years, indicative of an average recession rate of 0.31 feet per year, exactly one-tenth that of its eastern cousin.

Charleston Falls is small and unpretentious in comparison to American Falls, but because of similarities in rock type and rock age, they possess related life histories. Both are a product of the great Pleistocene ice ages, and both are doomed by the inevitability of falling-water erosion, which eventually leads to annihilation. Perhaps these are the reasons Charleston Falls has long been and rightly called a "miniature Niagara."

12. OAKES QUARRY
Sea-Lily Paradise

A Dayton-based building boom in the Roaring Twenties prompted a focused search for a new source of high-calcium limestone, an essential ingredient in the production of portland cement. Accessible exposures of 425-million-year-old Brassfield Limestone, 0.2 mile east of the intersection of I-675 and Ohio 235, fit the bill. The area became a productive quarry, which has since been transformed into the 190-acre Oakes Quarry Park, lauded as one of the largest and most diverse Silurian-age fauna sites in the world.

Oxygen-rich, warm, shallow-water seas, home to teeming masses of life, covered the park and its environs during the Silurian Period. Two coral reefs, exposed in the upper reaches of the quarry wall along the Fossil Trail, contain an amazing inventory of ancient marine organisms, but the presence of crinoids is exemplary, in terms of both degree of preservation and frequency.

Crinoids are relatives of the starfish, sea urchin, and sand dollar. These "lilies of the sea" have a long stem made of button-sized discs that are strung together like coins on a string. The calyx, composed of the vital organs, extends upward as a series of waving, feathery arms that constantly capture food particles. A rootlike holdfast anchors the animal to the seafloor. More than twenty-nine species of crinoid fossils have been found at Oakes Quarry Park, making this site a sea-lily paradise to both amateur and professional paleontologist.

The fossil-rich strata of the quarry wall represent a transgressive sequence of rock deposition, meaning the water became deeper with the passage of time. The uppermost exposures are beveled and grooved, evidence of glaciation that occurred more than 420 million years after the Brassfield Limestone was deposited. Fossil collecting is permitted in the sign-designated spoil-pile areas of the park, conveniently open every day of the year.

Signage along the high wall gives details of geologic history. —Courtesy of Kendall Hauer, Miami University

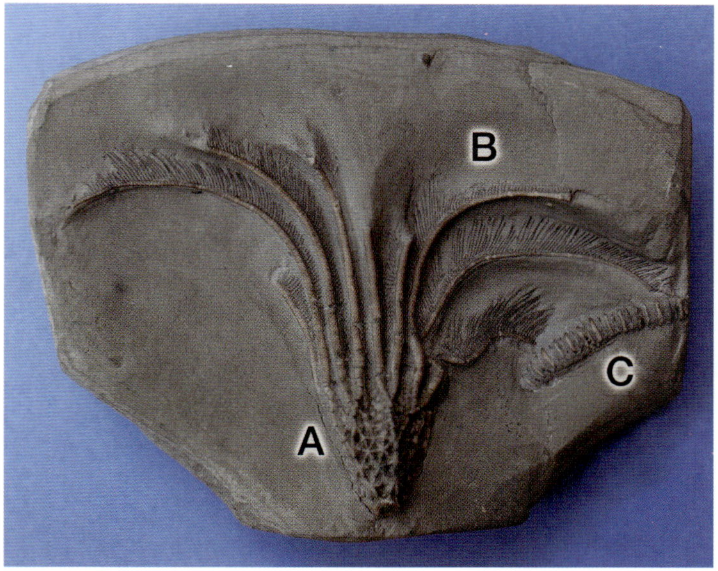

The three principle parts of a typical crinoid are seen in this 3-inch-wide fossil: the basal, basket-like calyx (A) extends upward to the remnants of nine feathery arms (B), and a detached portion of a stem (C) is to the right. —Courtesy of Kendall Hauer, Miami University

Bits and pieces of crinoids as well as whole specimens are preserved in this 28-inch-wide slab of limestone collected in southwestern Ohio. —Courtesy of Kendall Hauer, Miami University

Body parts of Silurian-age crinoids are scattered in this 4-inch-long, polished section of rock collected at Oakes Quarry Park. Dime for scale. —Courtesy of Kendall Hauer, Miami University

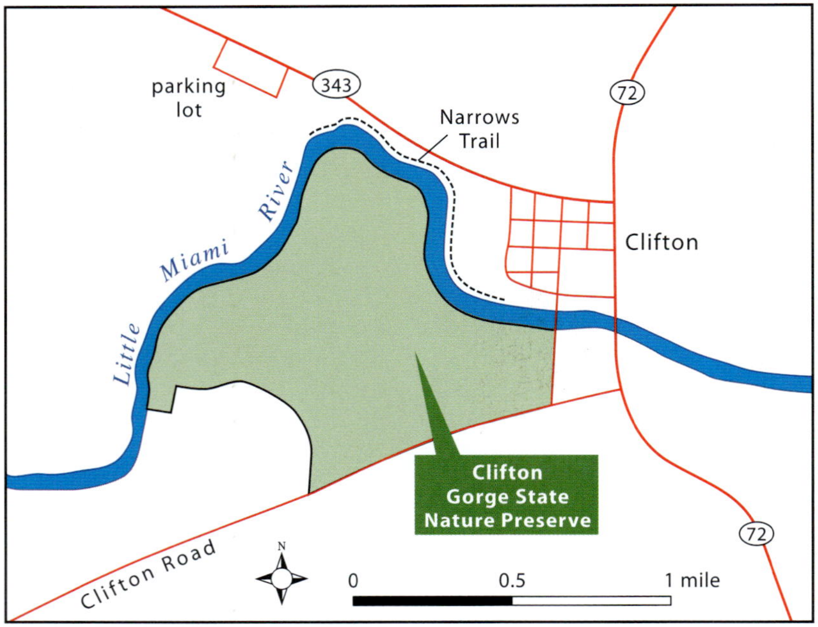

13. CLIFTON GORGE
A Chasm of Potholes

There is an estimated 332,500,000 cubic miles of water on and below the surface and within the atmosphere of Earth, and 96.5 percent of it is contained in the world's oceans. Far less that one-tenth of one percent is seen in the form of rivers, streams, and creeks. About 14,000 years ago, however, as the last remnant of the 1-mile-thick Pleistocene-age ice sheet retreated from Ohio, flowing water was the most important form of erosion crafting the landscape. Clifton Gorge State Nature Preserve, 0.5 mile northwest of Clifton off Ohio 343, is an outstanding example of one of these eroded landscapes.

Extensive layers of limestone and dolomite were deposited across the state some 430 million years ago by the Tippecanoe Sea, considered by many geologists the most extensive inland ocean to have ever covered North America. Thousands of millennia later, with the retreat of the last remnants of ice age glaciers from the state, the polished surface of these Silurian-age carbonate strata were deeply scoured by sediment-laden

Clifton Gorge is growing wider as less-resistant dolomite weathers into recesses, which causes the overlying resistant rock to collapse as slump blocks. —Courtesy of Rachael Garrity

Cascades of the Little Miami River separate walls of the massive Cedarville dolomite at Clifton Gorge.

meltwater. A myriad of potholes, sculpted by pebbles trapped in rotating eddies of fast-flowing water, enlarged and coalesced as a narrow, 50-to-100-foot-deep, 0.25-mile-long canyon that is lauded by many visitors as the most spectacular gorge in the Buckeye State.

This textbook example of postglacial canyon cutting can be examined up close along the Narrows Trail, which overlooks the eastern rim of the Little Miami River. The layers of rocks found in the depths of the gorge vary in their ability to combat erosion. While the resistant Cedarville dolomite identifies the upper portion of the steep-walled cliff face, less-resistant rock has weathered out beneath it to form deeply shaded recesses. This undercutting allows the Cedarville dolomite to separate along fractures and collapse to the valley floor in the form of massive slump blocks.

14. CIRCLEVILLE ESKER
A Glacial River Without a Valley

The concept was once quite controversial, but today it is an accepted fact that, multiple times in the geologic past, continental-sized glaciers covered some 70 percent of the land surface of Earth. The most recent occurrence that affected Ohio ended 14,000 years ago, when the last Pleistocene-age ice sheet was in full retreat after having enveloped two-thirds of the state in its frigid grip.

Evidence of glacial activity consists of topographic features created by both erosion and deposition. Of the many types of deposits attributed to retreating glaciers, eskers are both unusual and uncommon. Located in northern Pickaway County, the 10-mile-long, 20-to-25-foot-high Circleville Esker, defined as a fossil subglacial stream deposit that lacks an accompanying stream valley, is a beautiful example.

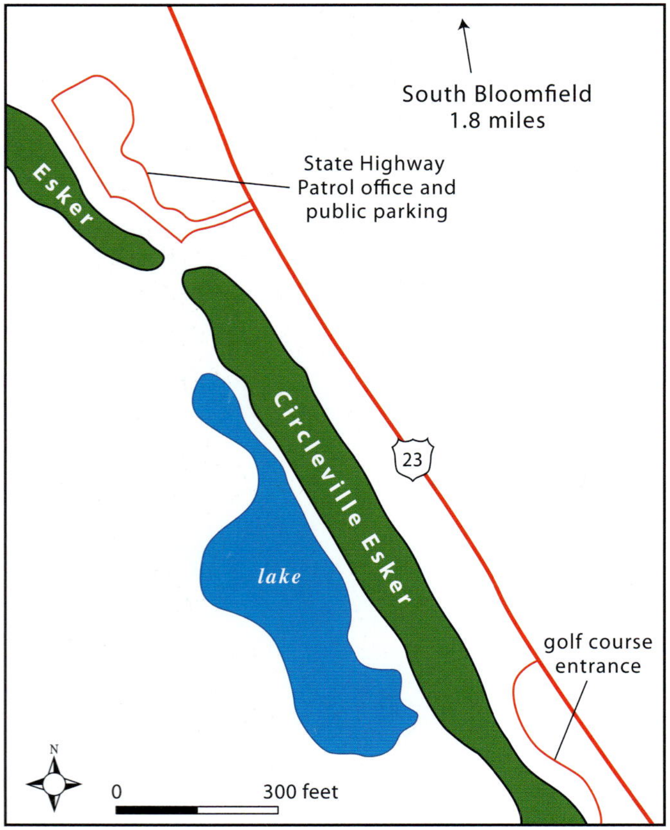

Looking south at the east slope (left) and axis (center) of the Circleville Esker from south of the State Highway Patrol office parking lot, 1.8 miles south of South Bloomfield.

In profile, the Circleville Esker is steep walled and rounded, with a convex-upward shape. Its base is flat and parallel to the surface on which it formed.

Glaciers reach the mature-citizen state when ice stagnates and begins to melt. Torrents of meltwater plunge vertically through fractures in the ice and then along narrow ice-walled tunnels at the base of the retreating glacier, depositing well-sorted gravel along the way. The resulting esker, left once the ice has melted completely, is a sinuous landform characterized as an elongate, steep-walled ridge oriented at a high angle to the glacier's former front. Because eskers contain valuable volumes of sand and gravel, they are frequent targets for commercial mining.

Studies suggest that in the not-too-distant geologic future the Buckeye State may once more be embraced by an expanse of ice more than 1 mile thick. In spite of today's global warming, these predictions, and the presence of the Circleville Esker, are evidence that the course of geologic history is subject to a tried-and-true corollary: what came around may once more come around.

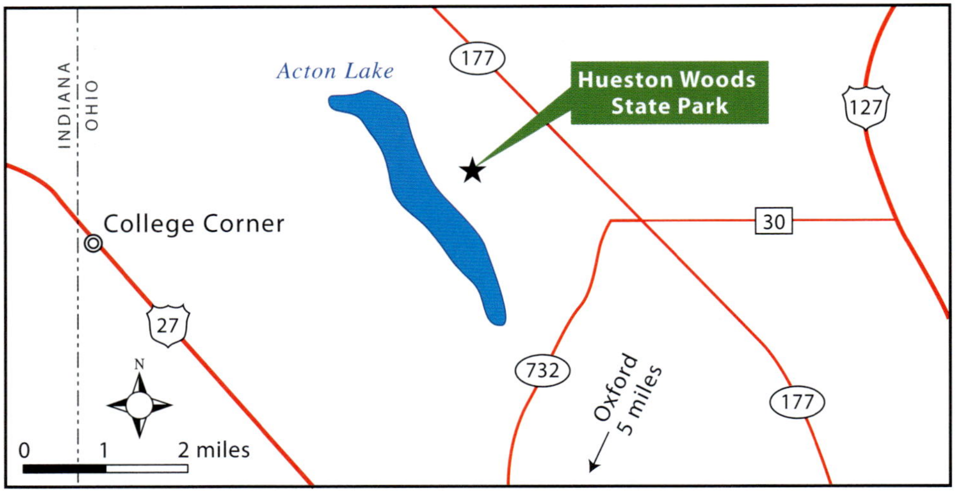

15. HUESTON WOODS
Fossils of International Fame

Late Ordovician strata of Ohio have an international reputation. Measuring 750 feet thick, these 460-million-year-old rocks comprise a mixture of interbedded limestone and shale that, at first inspection, appear monotonous and visually uninteresting. A closer look, however, reveals a treasure trove of ancient marine fossils cataloging more than 1,200 species. Indeed, the Cincinnatian Series strata (the formal name of Late Ordovician rocks in North America) are reputed to be some of the most fossiliferous found anywhere in the world.

The shallow-water sea that inundated central North America during the Ordovician Period is recognized by many as the greatest-ever submergence of any continent in all of geologic time. Shale-rich strata represent deposition in deep, quiet waters of this sea, while limestone-rich strata indicate deposition in clear, shallow water. Fossils abound in both rock types, but the wave action of ancient seas broke and abraded those preserved in limestone. Hueston Woods State Park in southern Preble County is a magnet for budding paleontologists drawn to the unusual beauty of intact Cincinnatian fossils. A map identifying recommended collecting sites is available at the park office.

Fossil ancestors to today's clam, squid, snail, and starfish families are present in quantity at Hueston Woods, but three categories are of special interest to the novice paleontologist because of their easy identification. Brachiopods—lampshells—are recognized by their bilateral symmetry: the left half and right half are mirror images of each other. Examples range from the half-dollar-sized *Rafinesquina* to the nickel-sized *Plaesiomys.* The solitary, extinct, several-inch-long coral *Grewingkia*, a rugose coral, is also sought after because of its size, wrinkled surface, and physical resemblance to a bull's horn. A sharp-eyed individual might even find a highly prized trilobite, the extinct, bottom-feeding king of the Ordovician sea. Good luck and happy hunting.

Common Hueston Woods brachiopods include 2-inch-wide specimens of the Rafinesquina *genus (top) and smaller, 1-inch-wide specimens of the* Plaesiomys *genus. These bottom-dwelling animals occupied large sections of the Ordovician sea in Ohio.* —Courtesy of Kendall Hauer, Miami University

Fossil corals of the Grewingkia *genus, both juvenile (bottom) and adult (top, 3.25 inches long), abound at Hueston Woods State Park.* —Courtesy of Kendall Hauer, Miami University

The highlight of any trip to Hueston Woods would be discovering a specimen of the elusive, 1-inch-long Flexicalymene *genus of trilobite.* —Courtesy of Kendall Hauer, Miami University

16. MIAMI BLUFFS
Three Tills and Some Loess

Today, populations around the world are being affected by unusual, and often adverse, weather conditions that an array of scientists attribute to global warming. This phenomenon of suspected atmospheric change is but one more account in a record of recognized climate variations dating back many millions of years. The most recent chapter with noticeable variation—the Pleistocene Epoch—began 2.6 million years ago, when an average worldwide temperature decline of as much as 12°F triggered the ice ages.

Ohio was then affected by a 1-mile-thick sheet of ice that advanced and retreated across the landscape during four extended, and numerous smaller, intervals of time. The most recent—dating from 24,000 to 14,000 years ago—and best preserved of the major invasions is known as the Wisconsinan stage. Sediments representing this stage are beautifully exposed along the Miami Bluffs in Peffer Park, adjacent to the campus of Miami University in Oxford.

Three examples of glacial till, each representing an episode of ice advance and retreat, overlie limestone and shale of the Ordovician-age Waynesville Formation, exposed immediately above the water level of Collins Creek. Largely covered by hillslope debris, the oldest till is associated with a 4-inch-thick, organic-rich layer of silt that formed in a marsh, as evidenced by fossil snail shells. The younger two tills are differentiated by slight variations in earth-tone color and subtle changes in the long-axis orientation of sediment particles that make up the till. Fine-grained layers of loess, silt, and sand—all glacial deposits transported by wind—form the upper several feet of Miami Bluffs.

As Pleistocene ice advanced across Ohio, many trees were plowed over and incorporated into the glacial debris. The glacial soils of southwestern Ohio are world-famous because they contain abundant examples of fossilized limbs, twigs, and the bark of trees and remnants of other plants that can be age dated using carbon 14 analysis, the technique used to date deposits less than 40,000 years in age. This process indicates the glacial tills of the Miami Bluffs are approximately 24,000 years old.

The nearby Limper Geology Museum, in Shideler Hall, is open to the public Monday through Friday. Home to 150,000 fossil specimens and more than 3,000 rock types, its thirty major displays beautifully supplement the geology of Miami Bluffs.

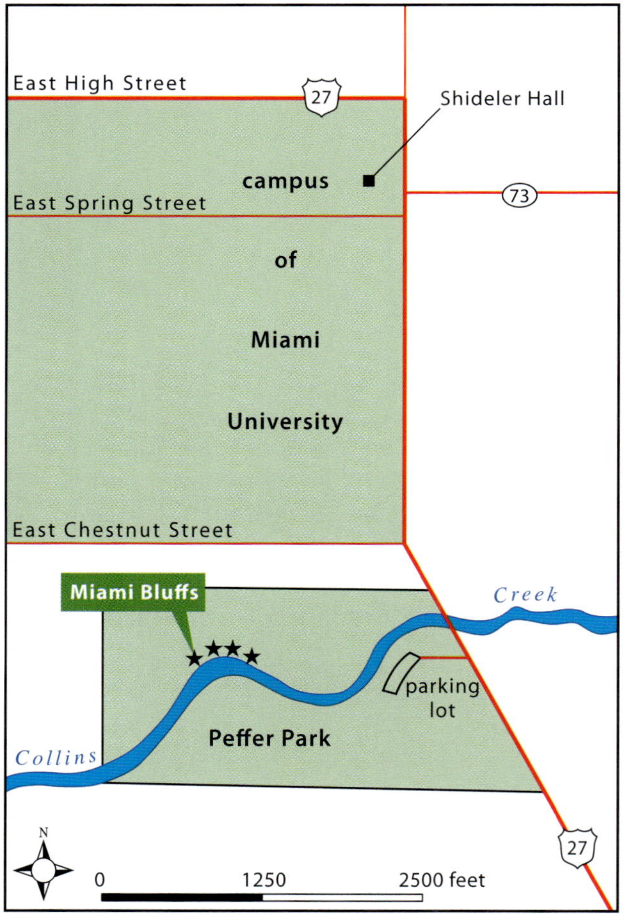

A 6-inch-long, 24,000-year-old spruce specimen collected from Miami Bluffs.
—Courtesy of Kendall Hauer, Miami University

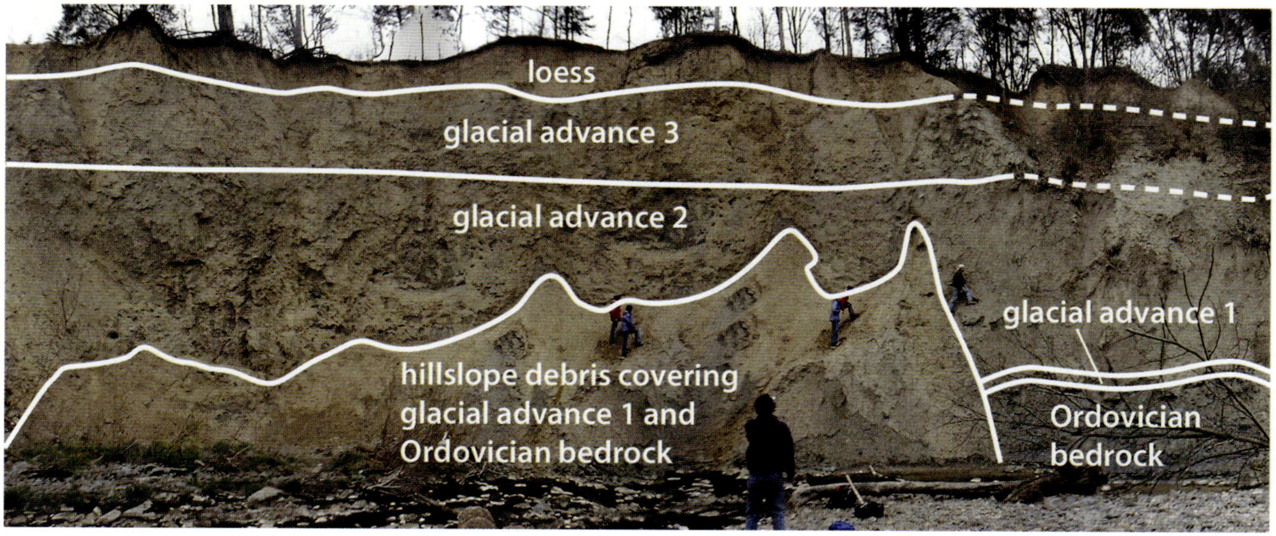

Subtle color changes help differentiate bedrock (lower right) from the hillslope debris, exposed glacial tills (marked 2 and 3), and overlying loess in the distinctive Miami Bluffs exposure. (Interpretation modified after M. P. Ekberg et al., 1993.)
—Courtesy of Kendall Hauer, Miami University

17. CAESAR CREEK
A Fossiliferous Bonanza

As the sun dawned on the first week of the Late Ordovician Period, around 458 million years ago, the land that would one day become Ohio occupied a venue far different from what exists today. Not only did it lie 15 degrees south of the equator, an extensive shallow-water ocean bathed by a subtropical climate also inundated the entire area. Under these benign conditions, multitudes of life proliferated within a saltwater environment similar to today's Caribbean Sea. Volumes of calcareous skeletons settled on the ocean floor, creating a world-class depository of fossils. Caesar Creek State Park is the site of choice for collecting fossils of this period of time. Brachiopods (bivalve creatures) are the most common, but plenty of bryozoans (colonial, mosslike animals) and corals exist as well.

With permit and map in hand, obtained without charge at the visitor center, one can collect a variety of palm-sized fossils at the nearby emergency spillway. Beautifully exposed and accessible here are three sequences of limestone and shale strata, named, from oldest to youngest, the Waynesville, Liberty, and Whitewater Formations.

Four species of coral, nine species of bryozoans, and twenty-four species of brachiopods can be found. The prima donna billing, however, belongs to *Isotelus*, a trilobite designated the official state invertebrate fossil of Ohio in 1985. Cousins to today's crabs, lobsters, and spiders, trilobites burrowed through seafloor sediment and obtained nutrients by ingesting mud. They once dominated the top of the food chain but became extinct 252 million years ago. Coiled specimens the size of a large marble commonly weather out of the shale beds. The rarely found prone specimens must be carefully excavated from fresh, unweathered shale due to their fragile nature. A truly outstanding 12-inch-long, prone example, discovered at Caesar Creek, is on prominent display at the visitor center. All trilobites grew by periodically molting their exoskeleton, a hard, outer shell similar in composition to the human fingernail. Thus, one trilobite could contribute numerous exoskeletons to the fossil record.

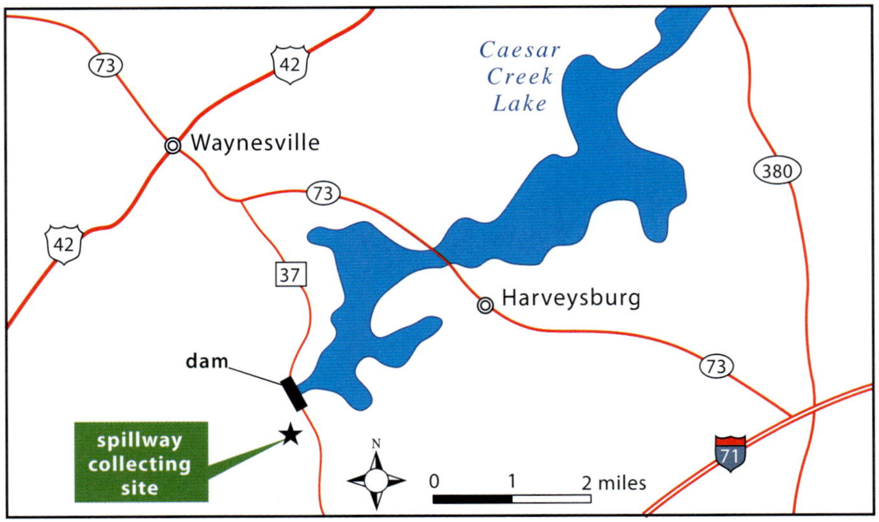

A perfectly preserved 1-foot-long specimen of the trilobite Isotelus. *—Courtesy of the Ohio Department of Natural Resources, Division of Geological Survey*

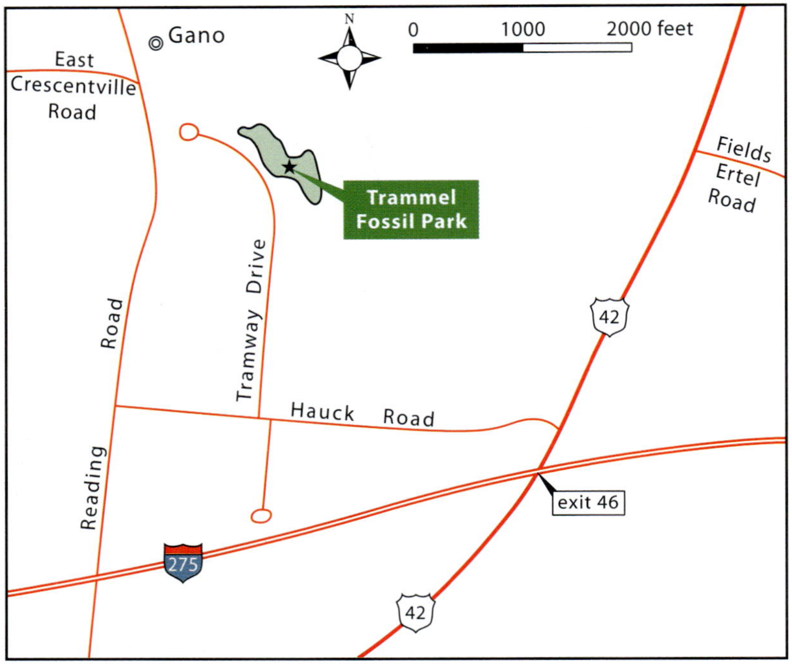

18. TRAMMEL FOSSIL PARK
When Cincinnati Was the Caribbean

Cincinnati is famous for several reasons: once the country's chief hog-packing center, it later became an important station on the Underground Railroad and then home to the Cincinnati Red Stockings, the first professional baseball team. Three defensible claims to fame—but there's another. The city is built over one of the most fossil-rich regions in North America, if not the entire world. Specimens of hundreds of Ordovician-age species, discovered in the limestone and shale formations exposed throughout the region, grace the display cases of paleontology museums across the globe. In short, Ordovician fossils are ubiquitous in and around Cincinnati. If your goal is to discover the *oldest* Ohio Ordovician fossils, then Trammel Fossil Park, north of exit 46 off I-275, is the site to visit.

Four rock units (designated by signs) are exposed at Trammel Fossil Park. The Fairview, seen at the base of the slope, is the oldest publicly accessible formation in Ohio. Shale and limestone of the Miamitown Shale, Bellevue Formation, and Corryville Formation overlie it.

Four formations, representing 2 million years of time, are exposed across the slopes of this 10-acre park. Numerous interpretative signs explain the geology at each of several distinctive levels. Because the older rock at the base of the slope contains more shale layers than limestone, geologists surmise it was deposited in a calm, relatively deep marine environment. Fine-grained sediment, such as that composing shale, can be carried long distances from a shoreline, finally settling out in calm water. In contrast, the limestone-rich rocks near the top represent a wave-agitated, shallow-water environment because limestone generally does not form in deep water. Specimens collected at the base of the slope are, therefore, apt to be found intact, whereas those found upslope are commonly broken into fragments.

This plethora of fossils tells the story of a portion of proto–North America 450 million years ago, when the Cincinnati region lay 15 degrees south of the equator in the embrace of a Caribbean-like climate. The fossils of Trammel Fossil Park, free for the taking, constitute a window into this distant chapter of Ohio history, when the environmental tempo was benign and absent of geologic catastrophe.

Fossil specimens found at Trammel Fossil Park include the 3.25-inch-long bryozoan Constellaria *(above), a small aquatic animal that reproduced by budding, forming mosslike, branching colonies, and three types of brachiopod (below), a solitary marine bivalve (front and back views shown).* —Courtesy of Kendall Hauer, Miami University

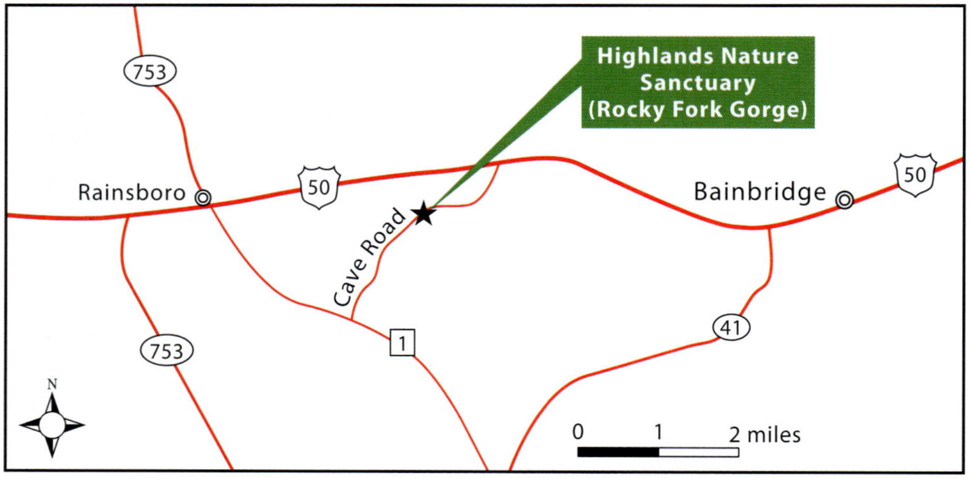

19. ROCKY FORK GORGE
Cornucopia of Karst

Many terrains compose the topographic profile of Ohio. Perhaps the most intriguing is karst, an undulating landform associated with carbonate bedrock. Formed when naturally occurring acidic solutions dissolve limestone and dolomite, karst is recognized by the presence of sinkholes, caves, and subsurface drainage.

Nestled deep in the old-growth forests of Highlands Nature Sanctuary, the 100-foot-high walls of Rocky Fork Gorge stand tall, sculpted into a foundation of Peebles Dolomite. Two scenic trails, accessed from the Appalachian Forest Museum, allow easy entry to a profusion of karst-developed features highlighted by a series of caves.

Etawah Woods Trail ends at the Three Sisters, a cluster of massive, upright slump blocks torn from the adjacent canyon wall by freezing water and gravity. Water froze in cracks along the canyon wall, expanding and wedging the blocks of rock loose; eventually, they could no longer defy gravity's tug. The midpoint of the Valley of the Ancients Trail is marked by Cave Run, a tributary of Rocky Fork Creek that flows along the floor of a cavern whose roof long ago collapsed.

The Peebles Dolomite was deposited as a fossil-rich limestone across the floor of a sea that flooded Ohio during the Silurian Period. Interaction with mineral-rich groundwater changed its chemistry to dolomite, altered limestone rich in magnesium. During this time of transition most of the fossils it contained dissolved, leaving the rock pockmarked with ghostlike cavities that mimic the shape and size of the destroyed fossils.

As the Pleistocene ice ages ended in Ohio, the loss of the weight of the melting glaciers caused the Peebles Dolomite to expand, and volumes of meltwater percolated through the developing vertical joints. At this time, some 14,000 years ago, the stage was set for the creation of the wondrous karst terrain of Rocky Fork Gorge, lauded as the second-densest concentration of caves in all of Ohio.

This pockmarked overhang of Peebles Dolomite is typical of the karst complexion in Rocky Fork Gorge. —Courtesy of Highlands Nature Sanctuary

20. SERPENT MOUND
Two-For-One Puzzle

Nestled on the edge of the Appalachian Plateau in northeast Adams County is a region of controversy and disorder associated with mysteries both terrestrial and extraterrestrial. Located off Ohio 73 between Locust Grove and Louden, Serpent Mound, the advertised attraction, is an imposing, convoluted sculpture winding 1,348 feet across the landscape in the shape of an open-mouthed serpent. Radiocarbon dating indicates that Serpent Mound, acclaimed as the largest prehistoric effigy mound in the world, was constructed around AD 1070, though there is still considerable debate concerning who did the building.

This singular, striated, cone-shaped, 7-inch-high shatter cone comes from Serpent Mound. —Courtesy of Kendall Hauer, Miami University

Few visitors realize that the coiled creature is poised on the western flank of a 4-mile-diameter circle of chaotic—folded and faulted—sedimentary rock that lies in sharp contrast to surrounding flat-lying strata. Discovered in 1838, the area of contorted rock was named Sunken Mountain, a moniker erroneously suggesting birth by compressive tectonic forces similar to those presently forming the Himalayas.

By 1920, scientists had come to favor a cryptovolcanic origin instead. Though they could find no direct evidence of volcanism, they envisioned a deep-seated chamber of magma vomiting gases to the surface with force sufficient to disrupt a regional volume of rock. Today, however, the mapped presence of shatter cones indicates beyond doubt that this geologic feature—officially

Shatter cones are also found in nested, or composite, groupings. This 3-inch-wide example formed in Ordovician-age limestone. —Courtesy of Mark Wilson, College of Wooster

named the Serpent Mound Disturbance—is one of thirty proven meteor impact craters in the United States, and the only one indenting the surface of Ohio. Shatter cones are clusters of fan-shaped structures that form in rocks, generally limestone, as a result of high-pressure, high-velocity shock waves produced by an impacting meteorite.

Three zones define the geology of the Serpent Mound Disturbance. A central uplift identifies the sector of meteorite impact, the force of which shattered and depressed the underlying strata. Within a fraction of a second a topographic high formed as the rocks rebounded. In compensation for this rise, an enveloping topographic low, the ring trough, formed during the same instant. The intervening region is the transition zone. Magnetic analyses suggest the impact occurred at the end of the Paleozoic Era, the time when the supercontinent Pangaea was close to being completely assembled, dinosaurs were about to begin their reign of terror, and the Atlantic Ocean was yet to be born.

Visual evidence of the Serpent Mound Disturbance is largely hidden from view by secondary forest growth. The contrast between its 4-mile diameter and the 1,348-foot-long Serpent Mound effigy—a magnitude ratio of almost 16 to 1—gives ample credence to the oft-stated claim that the Serpent Mound Disturbance is the largest, single-most-complex geologic structure exposed at the surface in Ohio. Little did the mound builders of yesteryear realize that their serpent-shaped monument marks the site of an ancient collision between a rocky space alien and Earth.

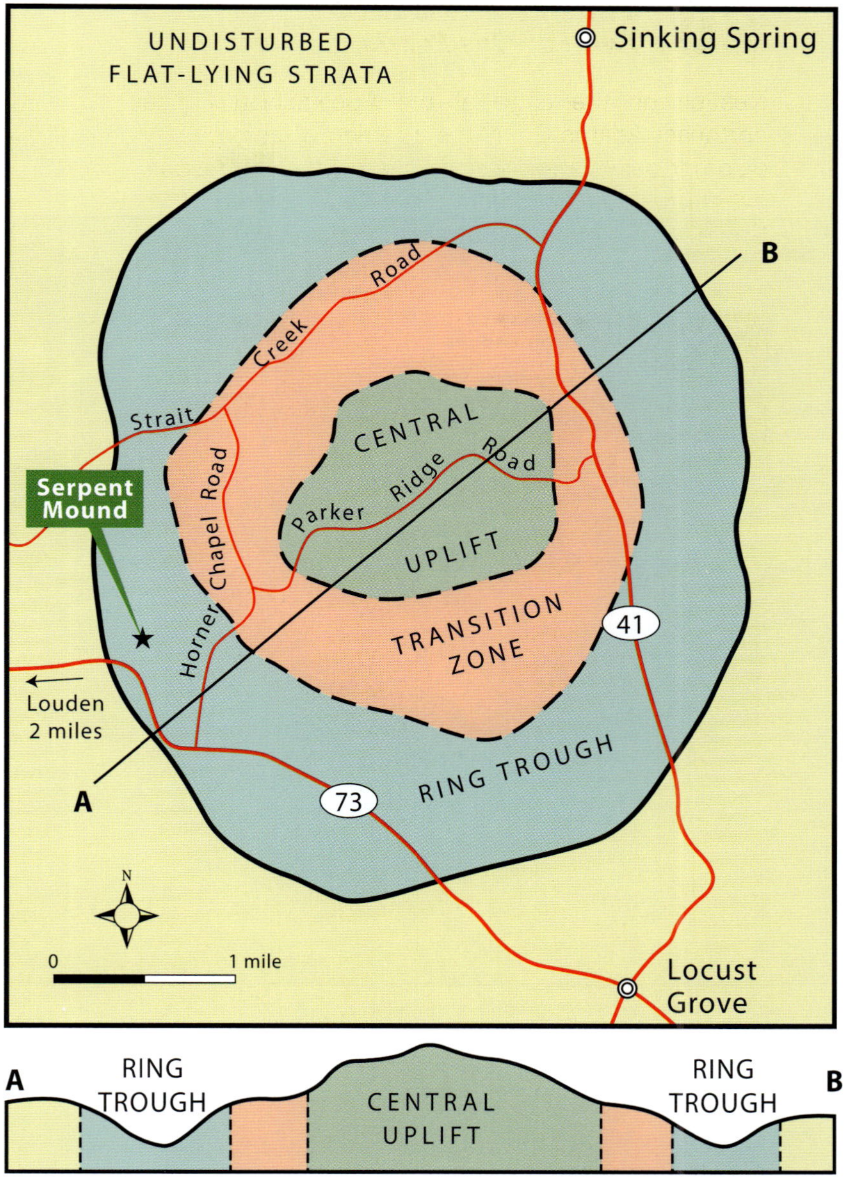

Geology of the Serpent Mound Disturbance. Note the 4-mile-diameter circular pattern caused by meteorite impact (heavy black line) and the subdivision into three topographic areas, shown in both map view and profile.

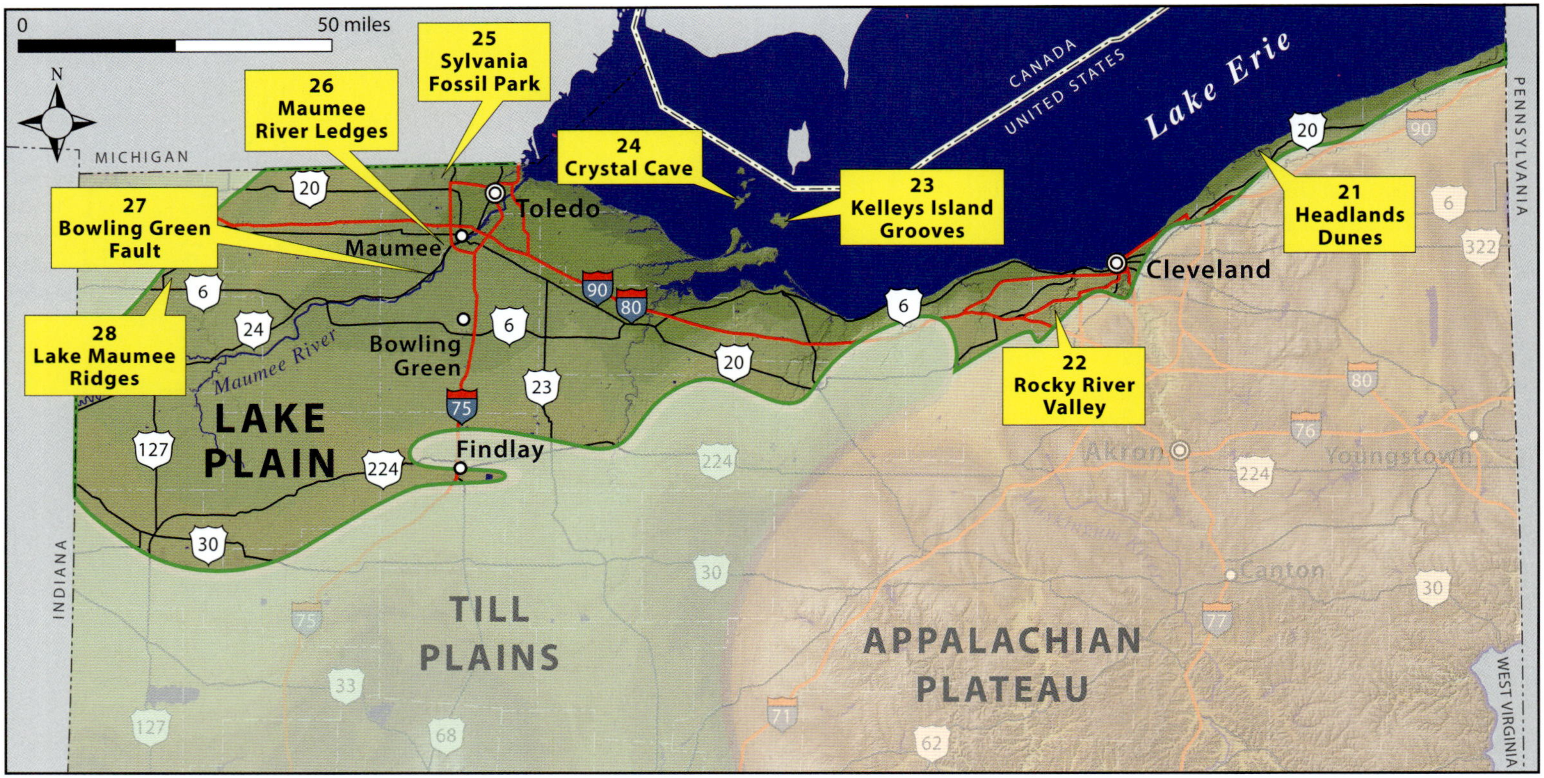

0
50 miles
N
MICHIGAN
INDIANA
PENNSYLVANIA
WEST VIRGINIA
CANADA
UNITED STATES
Lake Erie
21
Headlands
Dunes
22
Rocky River
Valley
23
Kelleys Island
Grooves
24
Crystal Cave
25
Sylvania
Fossil Park
26
Maumee
River Ledges
27
Bowling Green
Fault
28
Lake Maumee
Ridges
Toledo
Maumee
Bowling
Green
Findlay
Cleveland
Akron
Youngstown
Canton
Maumee River
LAKE
PLAIN
TILL
PLAINS
APPALACHIAN
PLATEAU
20
6
24
127
224
30
75
90
80
23
322
76
77
33
68
71
62

LAKE PLAIN

Prior to the beginning of the Pleistocene ice ages, 2.6 million years ago, the north-central portion of Ohio was drained by a river system that flowed to the northeast through basins now occupied by Lakes Erie and Ontario. As the Pleistocene wore on, the Erigan River, the master river in this system, as well as two-thirds of Ohio, was affected by a 1-mile-thick wall of ice that migrated back and forth across the landscape.

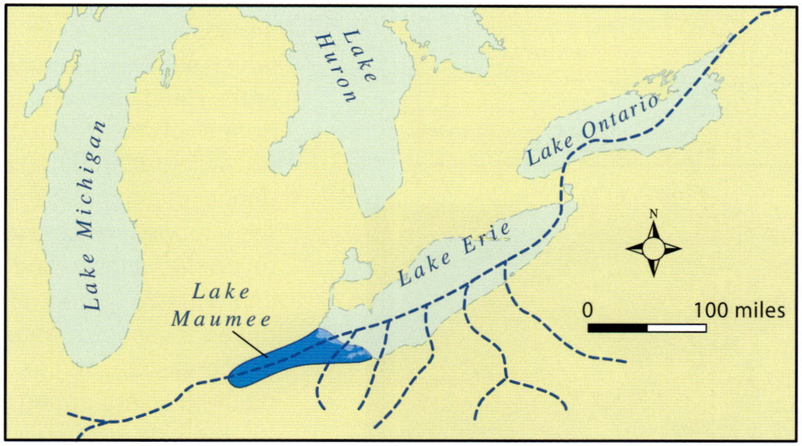

The preglacial route of the Erigan River (dashed blue lines) and its tributaries. Lake Maumee was the precursor to the larger, modern Lake Erie. Maumee developed adjacent to the retreating ice sheet as the ice melted and left Ohio for the last time. Outlines of the Great Lakes are shown for context.

The last ice retreated to the north and out of Ohio starting around 14,000 years ago, during the Wisconsinan stage, releasing large volumes of meltwater that flooded the ice-sculpted Erigan River valley and fed a complex series of lakes characterized by fluctuating water levels. The first of these was Lake Maumee. Modern-day Lake Erie came into being 4,000 years ago, but its early, high water level is identified by the sandy soil and flat relief of the Lake Maumee Plain.

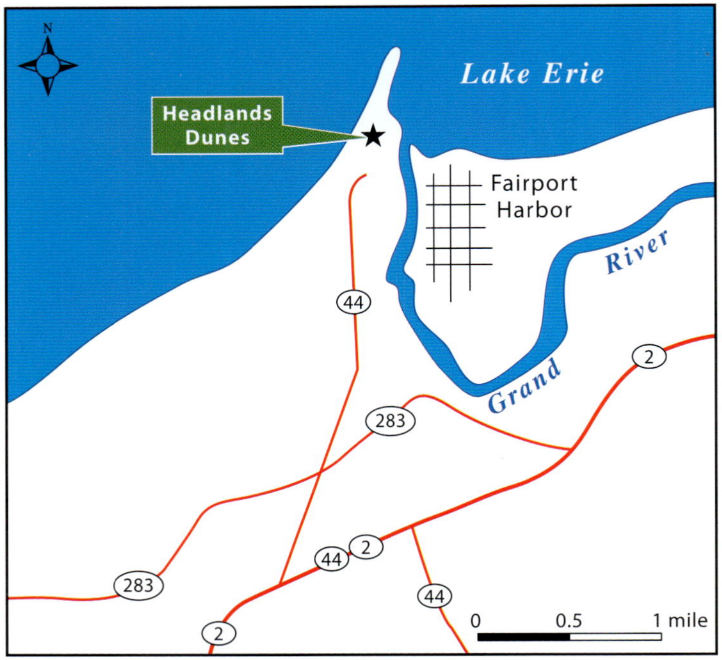

Silurian-age halite beds underlying Headlands Dunes have been mined since 1959 to supply ice-control salt for highways throughout the northeastern United States. —Courtesy of Tom Tomastik, Ohio Department of Natural Resources

21. HEADLANDS DUNES

Sand and Salt Aplenty

During the early years of the nineteenth century, northern Ohio settlers used the beaches of Lake Erie as a primary path of transportation. Lakeshore development has unfortunately eradicated most of this once majestic coastal byway, but an undisturbed and unusual example remains as the dune complex of Headlands Dunes State Nature Preserve, one of the last examples of this type of Lake Erie habitat.

Dunes generally develop where there is a sufficient supply of sand and strong winds. They form as blowing sand settles and accumulates around an obstruction, such as a clump of vegetation, which shelters the sand from the wind. The dunes at the Headlands preserve are transverse dunes—ridges of sand oriented perpendicular to the prevailing wind direction. A succession of plants define the dunes' ages: switchgrass anchoring the youngest dunes near the shoreline gives way to poplar, aspen, and eventually oak trees that characterize older inland dunes. These wind-driven lakeshore features are only one example of the vast resources of sand deposited in Ohio by the glaciers during the Pleistocene Epoch.

This site is closely associated with a subsurface natural resource. Since 1959, operations have been under way within a 2,000-foot-deep mine that directly underlies the Headlands preserve and extends for 2 miles under Lake Erie. The mine follows a massive, 25-foot-thick seam of Silurian-age halite (salt) that has been mapped across a 100,000-square-mile area, extending from Ontario east beyond central Pennsylvania and south into West Virginia. The product of the mine is 95 percent sodium chloride; the remaining 5 percent comprises other salts and rock impurities.

During Silurian time Ohio was located south of the equator and was inundated by a shallow, semi-isolated sea. High temperatures gradually evaporated the seawater, causing minerals to precipitate from it—first calcium carbonate and then halite. Discovery of this buried resource was strictly accidental, a happy if unexpected result of the drilling of an oil well in 1886.

Sand, like that forming the dunes at Headlands, and salt from an underlying fossil sea, are geologic products seldom given much thought, unless one is driving at freeway speeds along a windblown, snow-covered, icy highway. They then become products of unquestionable value.

The 750-foot walk from the Headlands Dunes parking lot to the shore of Lake Erie (over the horizon) is akin to a leisurely stroll through the geologic and biologic timeline of dune formation. —Courtesy of Jackie Oliverio

22. ROCKY RIVER VALLEY
Home to a Voracious, Toothless Killer

During Devonian time oceanic populations diversified to the extent that this time period is commonly referenced as the Age of Fishes. Primitive fish had been around for the better part of 100 million years, but during the Devonian sharks appeared on the scene, along with a group known as placoderms. Placoderms had a skeleton of cartilage and an armored head constructed around a massive toothless jaw. The biggest placoderm of all was *Dunkleosteus*, a monstrous, grotesque carnivore that quickly occupied the topmost position of the food chain.

Scientists have discovered fossil remains of this predator as far afield as Morocco, Belgium, and Poland. In Ohio, beautifully preserved examples are commonly found in the 370-million-year-old Cleveland Shale exposed along the valley walls of the Rocky River outside the John J. Donnelly Rocky River Nature Center, southeast of North Olmsted. Inside the center, a reconstructed *Dunkleosteus* skull is on display.

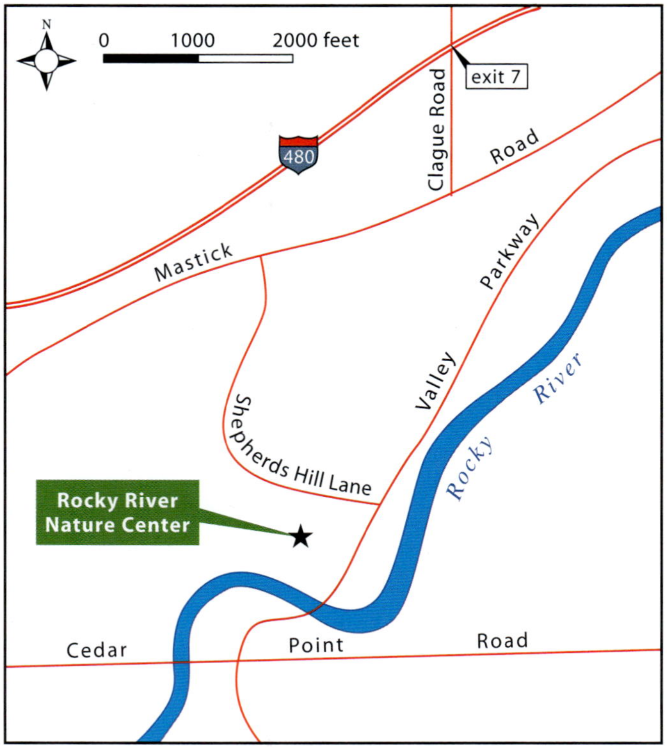

When fully developed, *Dunkleosteus* measured up to 35 feet in length and topped the scale at 4 tons. Biomechanical studies suggest it could open its bony-plated mouth in one-fiftieth of a second, causing a powerful suction that pulled in prey, and then close it with a bite that exerted a force of 80,000 pounds per square inch at the tip of its mouth, a bite that rivals that of modern alligators and *Tyrannosaurus rex*. Studies conducted on the sharp bony plates that formed its jaw imply juveniles had stiff jaws and fed on soft-bodied aquatic life, whereas the jaws of adults were more flexible in order to firmly grasp struggling prey. *Dunkleosteus* hunted anything that moved but had a particular appetite for sharks. The discovery of *Dunkleosteus* armor encrusted with bite marks indicates these formidable vertebrates would, on occasion, cannibalize each other.

The placoderm class of fish lasted a mere 50 million years. Compare this with sharks, which have been around for more than 400 million years. The extinction of *Dunkleosteus* was perhaps quickened by a tendency to suffer from indigestion: fossils are often found associated with masses of the semidigested and regurgitated remains of prey. Whatever the cause of their demise, a paraphrase of the old adage comes readily to mind: the bigger they are, the harder—and the quicker—they fall.

Four-foot-long reconstructed Dunkleosteus *jaw.* —Drawing by Patsy Faires

During the Devonian Period, Dunkleosteus *had a worldwide distribution and was the undisputed king of the predators. This statue is outside the nature center.*

The Rocky River has been eroding the Cleveland Shale since the end of the ice ages, opening the door to more discoveries of Dunkleosteus *fossils.*

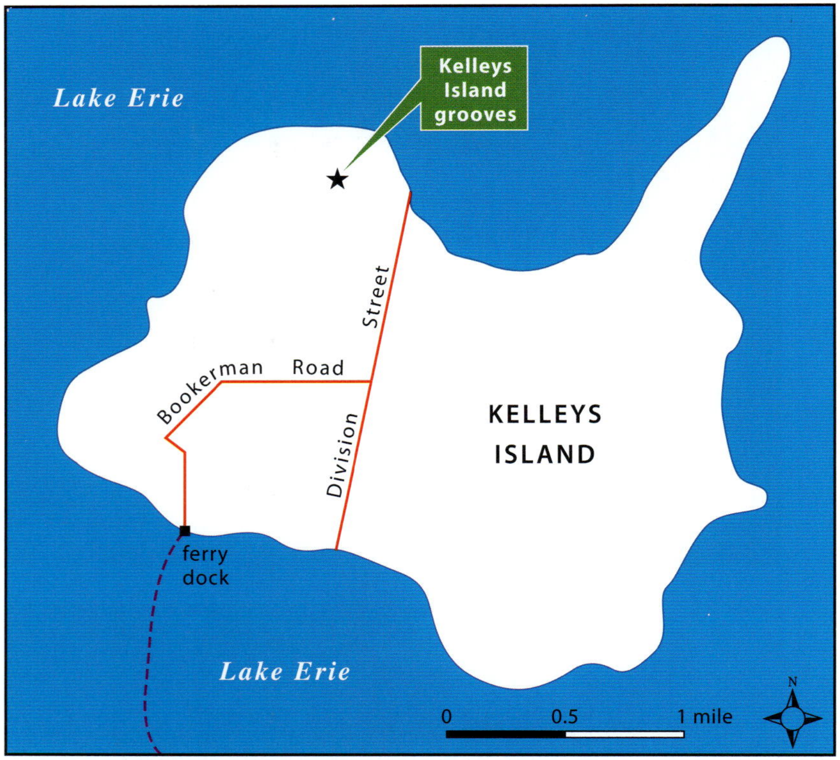

23. KELLEYS ISLAND GROOVES
Exquisite Rock Tattoos

Geologic history is rampant with controversy, the continuing give-and-take of sense and nonsense that is necessary in the establishment of truth. The subject of glaciation is a case in point. When it was proposed in 1760, many scientists roundly ridiculed the concept of ice in motion, but opinion changed in the nineteenth century when noted European geologist Louis Agassiz suggested a mobile blanket of ice, similar to the one that envelops Greenland, once draped Switzerland. American geologists listened and turned their attention to the gargantuan grooves etched into the limestone face of Kelleys Island in Lake Erie.

The surface of these megagrooves, semicircular depressions 30 feet wide and 15 feet deep, are scarred by shallow minigrooves that create a tattooed rock complexion comparable in appearance to the linear scratches coarse sandpaper leaves across a pine plank. There is little doubt that the motion of a great mass of ice, contaminated with multitudes of gritty sediment, sculpted the minute, needle-to-inch-width striations, or scratches, but the origin

The first-order, fractured, big-boy grooves deeply scoured into Devonian-age limestone bedrock at Glacial Grooves State Memorial. —Courtesy of the Ohio Department of Natural Resources, Division of Geological Survey

of the monster grooves is a matter of continuing debate. Glacially formed, yes, but the devil is in the details.

Throughout the Pleistocene Epoch, beginning some 2.6 million years ago, a 1-mile-thick ice sheet flowed back and forth from Canada into Ohio, exiting the state for good 14,000 years ago. Sometime during this interval of time the Kelleys Island megagrooves came into being. They were gouged out by one of three erosive forces: sediment-laden meltwater flowing beneath the ice sheet, boulder-riddled ice that deeply scoured the underlying limestone bedrock, or rock-encrusted ice that deepened preexisting river channels. The scientific jury remains confused and indecisive.

Controversy aside, these not-found-elsewhere, oversized grooves are a lasting and visual monument to moving ice's ability to carve and furrow Earth's bedrock.

24. CRYSTAL CAVE
World's Largest Geode

Some five hundred caverns undermine the integrity of the bedrock of Ohio. A dozen or so of these, holed into Silurian- and Devonian-age rock, fall along a north-south trend that extends from the caves of Highland County, east of Cincinnati, to the caverns of South Bass Island in Lake Erie.

A varied collection of sinkholes pimple the landscape of South Bass Island; they are a surface expression of the more than fifty caverns that riddle the subsurface. The subterranean caverns, comparable to the plethora of holes found in Swiss cheese, were all created by the same process: rather than being formed by dissolution

Crystal Cave. —*Courtesy of Analogue Kid at en.wikipedia*

Large and well-developed, bluish-white, translucent, tabular crystals are the principal reason Crystal Cave is truly one of a kind. —Courtesy of Kira Hubner, Crystal Cave

of carbonate bedrock, the common process whereby caverns develop, those of South Bass Island developed due to the dissolution of deep-seated layers of gypsum. This process created a void within the gypsum layer, into which the overlying dolomite collapsed.

Except for Crystal Cave, all of South Bass Island's caverns are ornamented with a collection of calcium carbonate stalagmites and stalactites. The dolomite walls of Crystal Cave, however, are embellished with large, morning-glory-blue crystals of celestite, an uncommon mineral formed of strontium sulfate. After the cave was discovered in 1897, miners removed many of the crystals, enlarging the cave. The strontium content of the crystals was valuable in the manufacture of fireworks. Fortunately, this practice was halted, and today this atypical site is open to the public daily for a small fee. The entrance is located 1 mile from Put-in-Bay, accessed by ferry either from Port Clinton or Catawba Island.

Migrating groundwater charged with sulfur and strontium precipitated the clusters of tabular-shaped celestite crystals, some more than 18 inches long, on the cavern walls. Similar occurrences of crystals exist on the island of Madagascar in the Indian Ocean, and in Argentina, but the exceptional number of crystals in the 30-foot-high Crystal Cave gives credence to its reputation as the world's largest geode.

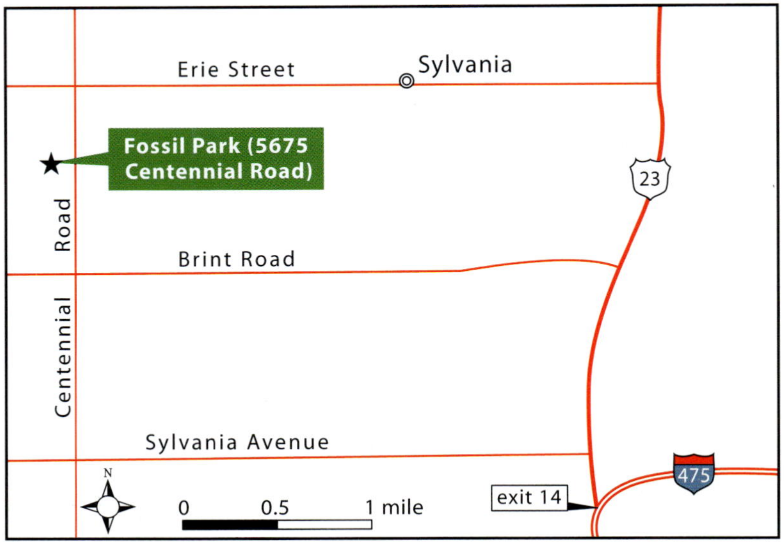

25. SYLVANIA FOSSIL PARK

Free Fossils!

For marine organisms, life during Middle Devonian time was good. Throughout this portion of Earth history, between 393 and 383 million years ago, Ohio lay 10 degrees south of the equator, thoroughly drenched by a shallow sea in which layers of limy sediment were deposited. The warm, transparent water teemed with a rich variety of invertebrate life: coral, clam, snail, mollusk, trilobite, and brachiopod. Primitive fish, some as big as sharks, the top of the food chain, filled their bellies with ease. Later, the water began to cloud as rivers flowing off the rising Acadian Mountains in the east carried massive volumes of silt and clay into Ohio. The environment was ideal for the formation and preservation of fossils as a large volume of sediment and skeletal remains of deceased organisms settled to the seafloor.

Over the following millions of years the sediments hardened into the Dundee Limestone and Silica Shale, formations world renowned for the fossils they contain. Rocks of this age are found in many places in Ohio, but one site merits special attention: Fossil Park in Sylvania.

This 5-acre abandoned quarry, publicized as one of only two prime Devonian fossil sites on the entire planet, is located north of the community of Silica. Admission is free, and so are the fossils. Periodically, drivers deliver fresh truckloads of Silica Shale—soft enough to be broken by hand—to the park from a nearby quarry. The budding paleontologist will have no trouble collecting a handful of exquisite specimens in the span of an hour.

Brachiopods, organisms that feed by filtering food from water, are especially common. The fossils of choice, however, are the harder to find horn coral and the large-eyed trilobite, the biggest prize of all. The horn coral was a solitary type of coral with a distinctive horn shape, and the trilobite was a bottom-dweller that fed as it cruised through the slime and mud of the ancient ocean floor. Both of these invertebrate organisms are extinct.

The Megastrophia *genus of brachiopod is a common fossil at Fossil Park.* —Courtesy of Kendall Hauer, Miami University

The rare and prized trilobite Phacops *found food with a pair of multifaceted compound eyes that had a wide range of vision.*
—Courtesy of Kendall Hauer, Miami University

The smaller Mucrospirifer *genus of brachiopod is also found at Fossil Park.*
—Courtesy of Kendall Hauer, Miami University

26. MAUMEE RIVER LEDGES
Alvar of Ancient Algae

One of the more challenging exercises associated with the practice of geology is the construction of paleogeographic maps, which are representations of the physical geography of particular locations in the geologic past. Interpreting the geology of Lucas County in order to create a map reflecting the middle epoch of the Silurian Period is a classic example of this difficult task.

The first-order environments that constitute global geography—land (desert, mountains, prairie) and ocean (tropical, deep, nearshore)—tend to be the easiest to reconstruct. The remaining major environment—the shoreline—is the most difficult to map, because it often includes a variable mixture of both land and sea ecologies: marsh, delta, tidal, brackish water, freshwater . . . and the list goes on.

A paleogeographic map representing the flat-lying rock ledges exposed along the north bank of the Maumee River in Side Cut Metropark would show circular, domed mats of algae scattered across a calcareous-rich, muddy tidal flat. The algae were a type of seaweed whose growth depended on nutrients supplied by daily tides. This environment characterized northern Ohio 420 million years ago, during Silurian time. A modern-day analogue to this ancient shoreline would be the coastal flats bordering the Bahamas, off the coast of Florida.

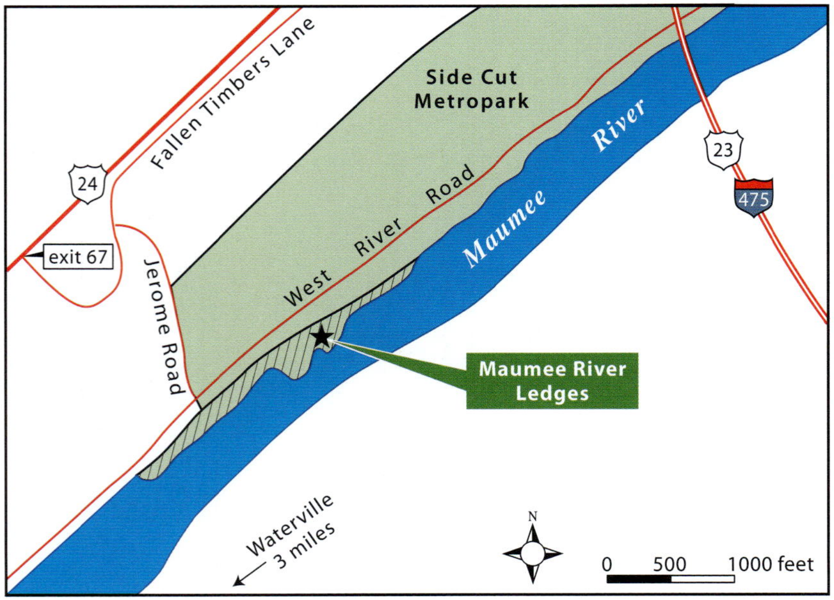

Telltale fossil evidence of this former shoreline includes mud cracks, an occasional ripple mark, and a wide spattering of laminated bedrock features—the fossilized algae—in the Maumee River. These circular features could be interpreted as evidence of ancient whirlpools or the fossil remains of clumps of flood debris. Microscopic analysis of a cross section of any set of the bonded layers that compose the mounds, however, indicates they are composed of alternating layers of organic cells and sediment. The fact that they are actually pods of fossilized algae shows the confusion geologists often encounter when attempting to create paleogeographic maps.

The Maumee River Ledges are considered an exceptional example of an alvar, a limestone plain covered with a thin layer of soil and little vegetation. More than one hundred alvars exist around the Great Lakes, but the Maumee River site is one of only three along the shore of Lake Erie. This arresting combination of fossil shoreline and modern-day alvar makes the Maumee River Ledges a geologic site of significance.

The tractor-tire-sized circular feature (center) is a pod of fossilized algae.

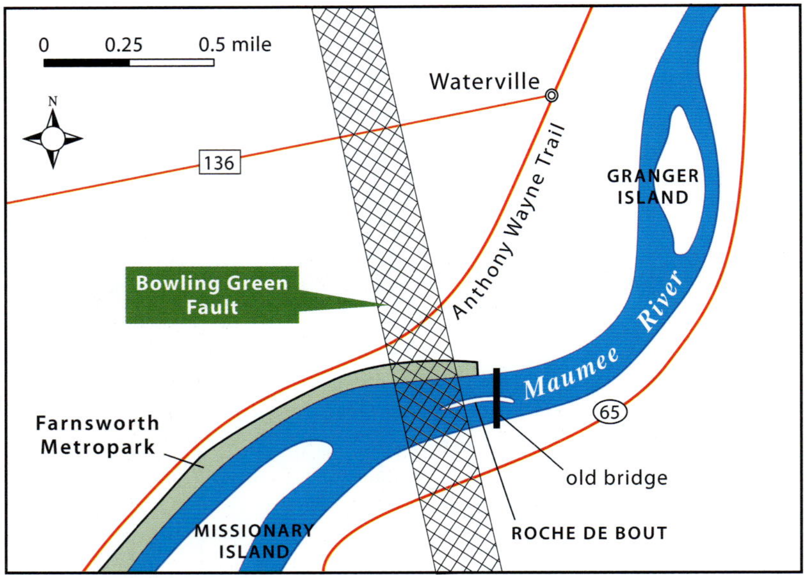

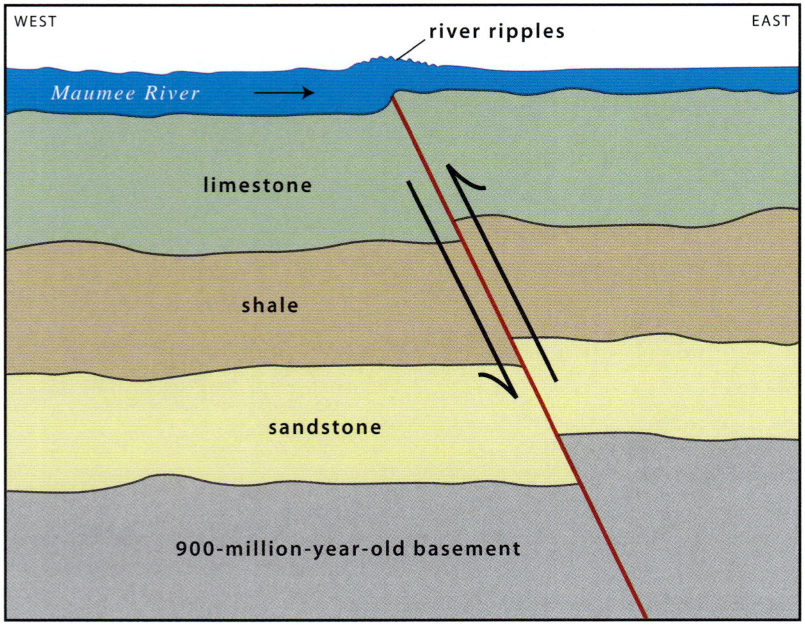

Cross section showing the Bowling Green Fault (red line) in Farnsworth Metropark. The arrows show the relative movement along the fault that caused a series of ancient earthquakes.

27. BOWLING GREEN FAULT
Antique Earthquake Zone

California is well-known for the numerous active faults that fracture its surface. These linear zones of rock under pressure frequently rupture, releasing spasms of energy that result in destruction and death. By measure of earthquake occurrence and intensity, California is a youthful state. In contrast, Ohio is a senior citizen. Its mere dozen or so documented faults are, with one exception, deeply buried in basement rock and rupture on a sporadic basis with intensities that seldom cause any degree of extensive damage.

The exception is the Bowling Green Fault, a north-south fracture that extends 100 miles from Michigan southward to the city of Findlay. Although it's mostly concealed beneath glacial sediment, a small segment is exposed in Farnsworth Metropark, near Waterville. The fault is recognized as the most significant structural geological feature of northwestern Ohio.

When two segments of Earth's crust were crushed together some 900 million years ago, the collision rearranged Ohio into a trend of complex terrain commonly referred to as the Grenville Mountains. Erosion long ago diminished the heights of these Precambrian-age mountains, making them part of Ohio's deep basement, but the Bowling Green Fault remains as evidence that the Buckeye State once sported an impressive alpine profile associated with a long history of earthquakes. Studies indicate the last earthquake produced by this fault occurred 100 million years ago.

When the Maumee River flows at a normal stage, the Bowling Green Fault creates ripples. The fault crosses the river immediately upstream and to the right of the Roche de Bout (center).

The ripples that disturb the flow of the Maumee River upstream of the historic Roche de Bout landmark, in the eastern section of Farnsworth Metropark, identify the fault where it crosses the river. When the water level is low, the trace of the fault across the bedrock is visible, physical testimony to that distant time when northwestern Ohio was shaken, rattled, and rolled in a manner similar to California today.

28. LAKE MAUMEE RIDGES
Shorelines of an Extinct Lake

Glacial activity can drastically transform the appearance of any area, as seen in northwestern Ohio. As a 1-mile-thick wall of ice moved back and forth across the state during the Pleistocene Epoch, the land was given a geographic face-lift. Old wrinkles were removed, leaving a flat profile, and a region once relatively high and dry was left low and wet. The ice bulldozed away layers of sedimentary rock, creating a gap in the text of geologic history.

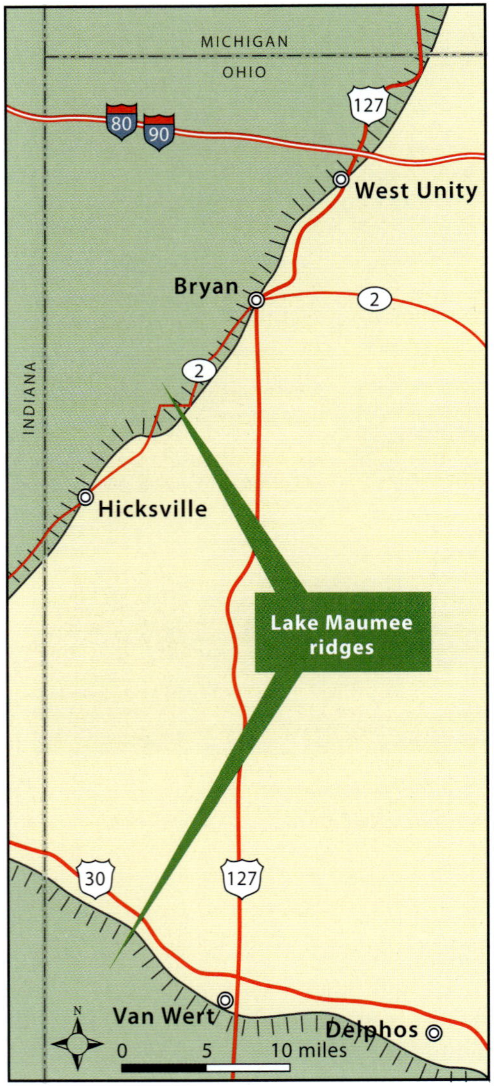

In northwest Ohio, beach ridges (dashed lines) separate clay and silt deposits (tan) of Lake Maumee from till deposits (green) left by the melting ice as it retreated to the north.

When glacial meltwater flooded a chain of low-lying basins 14,000 years ago, the Great Lakes were born, lapping at the front of the receding ice sheet. They were initially quite large, but as erosion opened new outlets, the lakes shrank. Scientists have carefully mapped and named the various shoreline positions of each Great Lake. The highest water level of Lake Erie—the twelfth largest freshwater lake in the world by area—is known as Lake Maumee.

Discontinuous ridges of sand and gravel found parallel to Ohio 2 between West Unity and Hicksville identify the northwestern shoreline of Lake Maumee, and similar ridges between Van Wert and Delphos along US 30 identify the southwestern shoreline. Due to their low relief, these high-water borders are difficult to see from ground level. Their geographic positions, however, are well-known through the mapping of the regional contact between sand and silt deposits, which contain freshwater fossils, and layers of glacial till, identified by the presence of terrestrial plant and animal fossils.

Standing on these Lake Maumee ridges, where till and lake sediment make contact, one can magically return to the last days of the most recent ice age, when the waters of the very youthful Lake Erie inundated the landscape as far to the southwest as Fort Wayne, Indiana. The extent of Lake Maumee was still visually evident in the mid-1800s in the form of the Great Black Swamp, a 1,500-square-mile area of wetlands and marshes that both the Native Americans and early settlers feared. Drained late in the nineteenth century, today the Great Black Swamp region contains some of the most fertile farmland in the United States.

The fine-grained, massive, yellow-toned Lake Maumee sediment (left) contrasts sharply with gray, crumbly, and coarse glacial till. —Courtesy of the Ohio Department of Natural Resources, Division of

Between West Unity and Bryan, US 127 follows the very crest of ancient Lake Maumee. Looking southwest, glacial lake deposits lie to the left, while terrain to the right is composed of glacial till.

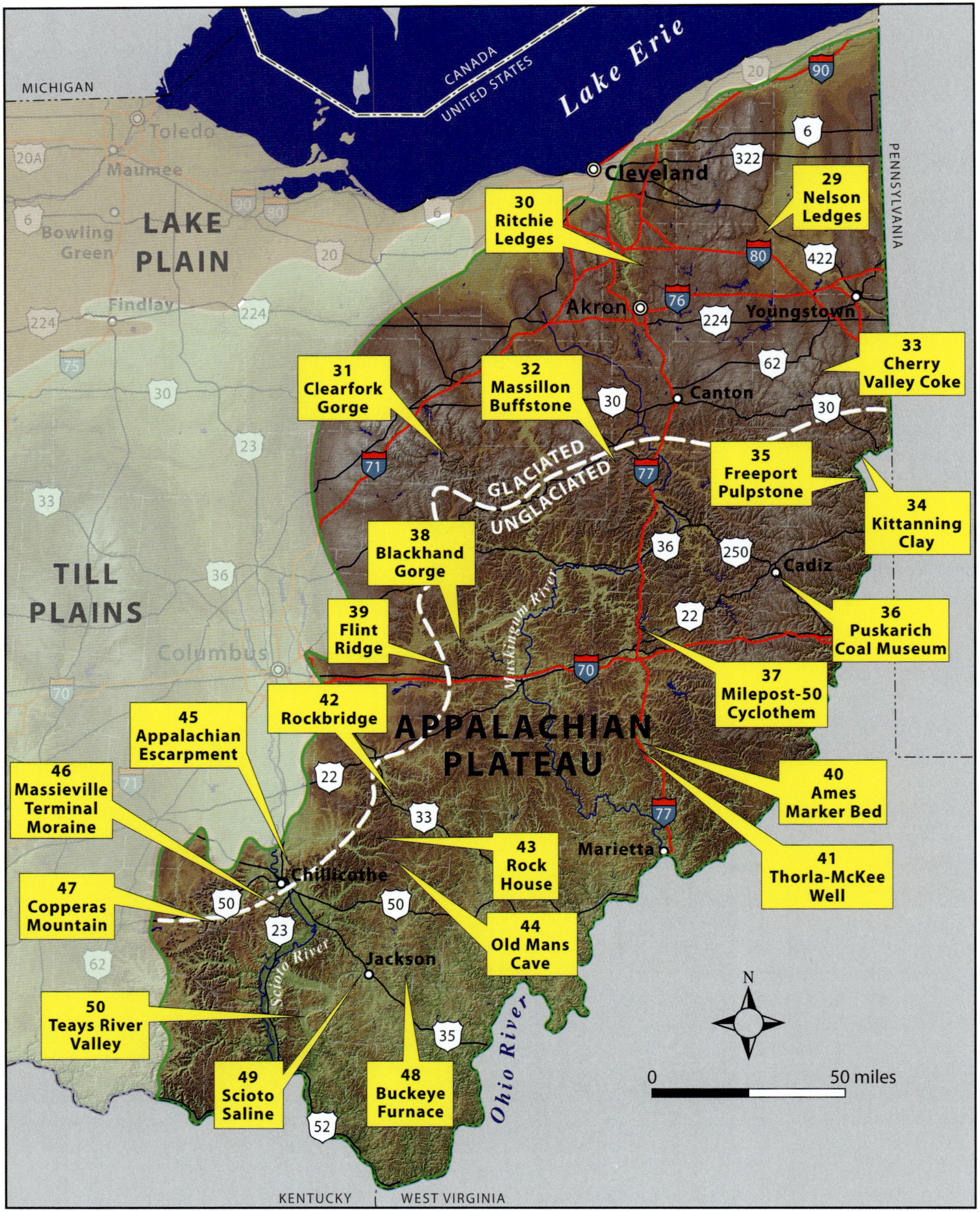

Lake Erie
CANADA
UNITED STATES
MICHIGAN
PENNSYLVANIA
KENTUCKY
WEST VIRGINIA
LAKE PLAIN
TILL PLAINS
APPALACHIAN PLATEAU
GLACIATED
UNGLACIATED
Toledo
Maumee
Bowling Green
Findlay
Columbus
Cleveland
Akron
Youngstown
Canton
Cadiz
Marietta
Chillicothe
Jackson
Muskingum River
Scioto River
Ohio River
29 Nelson Ledges
30 Ritchie Ledges
31 Clearfork Gorge
32 Massillon Buffstone
33 Cherry Valley Coke
34 Kittanning Clay
35 Freeport Pulpstone
36 Puskarich Coal Museum
37 Milepost-50 Cyclothem
38 Blackhand Gorge
39 Flint Ridge
40 Ames Marker Bed
41 Thorla-McKee Well
42 Rockbridge
43 Rock House
44 Old Mans Cave
45 Appalachian Escarpment
46 Massieville Terminal Moraine
47 Copperas Mountain
48 Buckeye Furnace
49 Scioto Saline
50 Teays River Valley
N
0
50 miles

APPALACHIAN PLATEAU

The Appalachian Plateau physiographic province occupies the eastern half of the Buckeye State and is composed of a glaciated and unglaciated region. Each region has a distinctive topographic profile.

Rolling hills and broad, flat valleys pockmarked with bogs and kettle lakes identify the glaciated region. In contrast, angular ridges constructed of erosion-resistant sandstone and youthful V-shaped hollows cut into less-resistant shale beds define the unglaciated area. Backcountry roads and twisting streams meander through this combination of high and low topography like strands of boiled spaghetti.

While the soils are at best low in fertility and acidic in nature, the Appalachian Plateau is rich in second-generation timber resources and harbors a world-class inventory of fossil fuels: bituminous coal, petroleum, and natural gas.

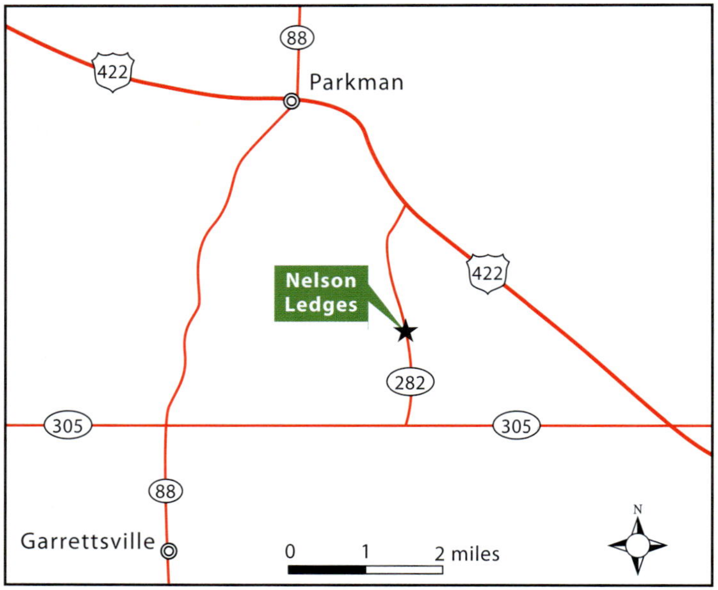

29. NELSON LEDGES
Sapping and Sliding

Many types of natural disasters affect humanity, with floods, tornadoes, and earthquakes receiving much of the publicity. While the role of mass wasting is seldom mentioned, it is responsible for twenty-five to fifty deaths and $1 to 2 billion in losses on average each year in the United States. Defined as the movement of bedrock, sediment, and soil under the influence of gravity, mass wasting is one of nature's ways of reducing every hill and mountain to a state of minimal relief. The imposing sandstone ramparts of Nelson-Kennedy Ledges State Park offer an unusual opportunity to observe the effects of this slow but persistent process.

The cliffs of Nelson-Kennedy Ledges State Park are composed entirely of sedimentary rocks. Cuyahoga Shale, deposited during Mississippian time, lies beneath thick layers of Pennsylvanian-age Sharon Conglomerate, characterized by ubiquitous milky-white quartz pebbles that range from pea to pigeon egg in size. The contact of these units is exposed in the recesses of Gold Hunters Cave, underlying Cascade Falls.

Shale and sandstone differ in their resistance to erosion, with shale being more susceptible. The undermining of soft strata beneath the erosion-resistant Sharon Conglomerate, a process known as sapping, creates a debacle

The weathering of Sharon Conglomerate creates a forest floor littered with rounded, milky-white quartz pebbles. Quarter for scale.

ready for the happening. With loss of support, and aided by gravity and the presence of multiple sets of fractures, massive blocks of rock separate from the cliff face and slide downslope.

The displaced blocks will eventually be reduced to scattered patches of sand that meandering river systems will ultimately transport out of the region. The final destination, however, of any one sand grain is uncertain. Because Nelson-Kennedy Ledges State Park sits atop a continental divide, some grains will eventually end up in the Gulf of Mexico, while others will find their way to the mouth of the St. Lawrence River, which will deliver them to the Atlantic Ocean.

Cascade Falls and Gold Hunters Cave (center), viewed from the safety of the observation platform. —Courtesy of Michael Flavin

Sunlit paths meander between massive blocks of displaced Sharon Conglomerate (left) that separated from parent bedrock (right) through the process of sapping and sliding.

30. RITCHIE LEDGES

Braids and Honeycombs

The bedrock foundation of Cuyahoga Valley National Park, a 52-square-mile arena of forested landscape situated along the Cuyahoga River between Akron and Cleveland, is constructed of shale, sandstone, and conglomerate of the Carboniferous Period. An impressive array of steep ravines, meandering streams, and plunging waterfalls characterizes the park, but the brutish, fractured walls of Sharon Conglomerate in the Ritchie Ledges unit are otherworldly features of geologic attraction.

Crossbedding, the occasional ripple mark, and thin, irregular lenses of rounded, milky-white quartz grains are convincing evidence that the Sharon Conglomerate was deposited along the shoreline of a shallow sea. The sea in question invaded Ohio 320 million years ago. The quartz grains were supplied by a system of braided streams: tangled networks of converging and diverging channels separated by narrow sand and gravel islands.

Cliffs of Sharon Conglomerate adjacent to the 1.8-mile-long Ledges Trail that encircles Ritchie Ledges are tattooed with an uncommon form of rock decomposition: honeycomb weathering. Found throughout the world, and even observed in martian rocks, honeycomb weathering remains a subject of controversy and mystery. Defined as a process of weathering that cosmetically alters a rock face into a pitted surface resembling an enlarged honeycomb, it has been described also as "stone lattice" and "stone lace."

Numerous explanations for honeycomb weathering have been offered: wind erosion, temperature changes within the rock, and the effects of gelifraction, the mechanical disintegration of rock by the pressure created by water freezing within cracks and pores. Another hypothesis suggests that differential weathering is the answer. As pockets of limonite cement that bind the sand and pebble grains of the conglomerate together dissolve, the grains fall away and the pitted surface develops.

Whatever the cause, the honeycomb texture of the Sharon Conglomerate only adds to the allure of Ritchie Ledges, a rugged sanctuary of solitude, beauty, and diverse geologic processes.

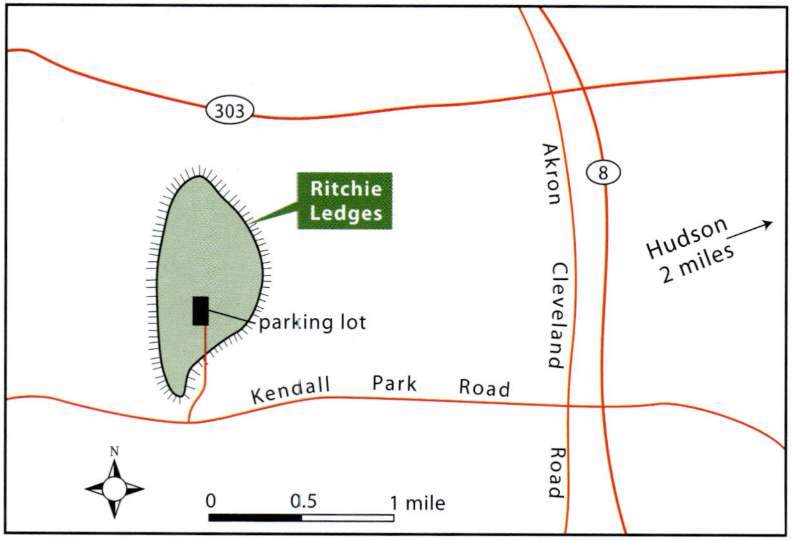

Ritchie Ledges rock faces altered by honeycomb weathering. Pen for scale.

Chevron crossbedding (center) overlain by regular crossbedding is evidence of a change in sediment deposition; the former was deposited under the influence of ebbing and flowing tidal currents, whereas the latter was deposited by a river flowing from right to left. —Courtesy of Michael Flavin

31. CLEARFORK GORGE
Opposing Rivers Become One

The massive rock exposures shaded in the sylvan setting of Mohican State Park harbor a record of two ancient invasions, one of a constructive nature and the other destructive. Together, the invasions relate a chapter of geologic history accentuating how the complexion of Mother Earth is constantly changing.

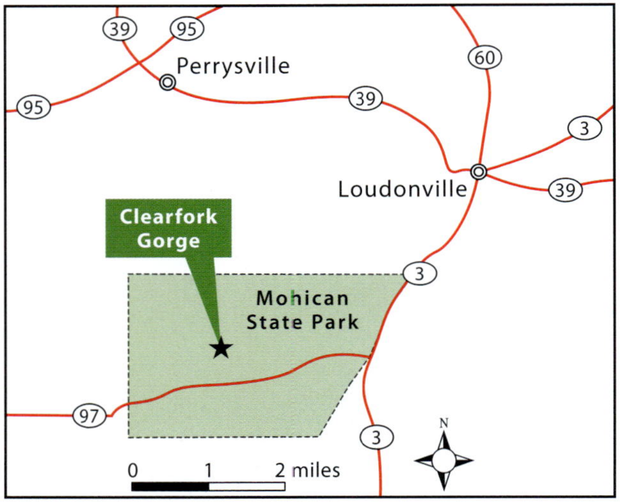

The constructive invasion began 350 million years ago, when shallow water inundated the land that is today Ohio. Rivers flowing off the rising Appalachian Mountains deposited sediment in the sea. With time and pressure, the sediments became stacks of sedimentary strata. The most impressive formation, constructed entirely of a massive cliff-forming rock, is the quartz-rich Black Hand Sandstone, which forms the bedrock backbone of Mohican State Park.

Fast-forward to 2 million years ago, when the same land was under attack. The aggressor was a 1-mile-thick, destructive column of ice that gouged deeply into the Black Hand Sandstone, creating a series of cliffs and rock shelters in Mohican State Park. Before the onset of glaciation, a north-south-oriented ridge, located in the center of the park, separated the local drainage system into streams that flowed in opposite directions: one to the east, the other to the west. When a tongue of ice dammed the west-flowing river, a lake formed. It increased in volume and elevation until it reached its spill point. The overflowing water cut a new channel across

A clutch of 4-to-5-foot-wide boulders is evidence that Big Lyons Falls is slowly being destroyed by rockfall, a type of erosion often caused by the freezing and thawing of water in cracks in the bedrock.

the ridge, joining the two valleys that lay on opposite sides of the ridge. After the ice dam melted, the newly born Mohican River flowed exclusively to the east through the 300-foot-deep, 1,000-foot-wide Clearfork Gorge, which is lauded as the deepest gorge in Ohio.

Black Hand Sandstone is beautifully exposed at Big Lyons Falls, a textbook example of a hanging valley, a tributary valley that is relatively higher than the valley it feeds into. During glaciation, a large glacier occupied the valley of the Mohican River, with smaller glaciers feeding into it from side valleys. Because of its enhanced ability to erode, the large glacier cut deeper into the bedrock than those in tributary valleys. When the ice melted, the mouths of the smaller tributaries were left high up on the steep side of the larger glacial valley.

Big Lyons Falls flows over an 80-foot-high fractured wall of Black Hand Sandstone. The falls is a hanging valley. —Courtesy of Patty Rickard, University of Mount Union

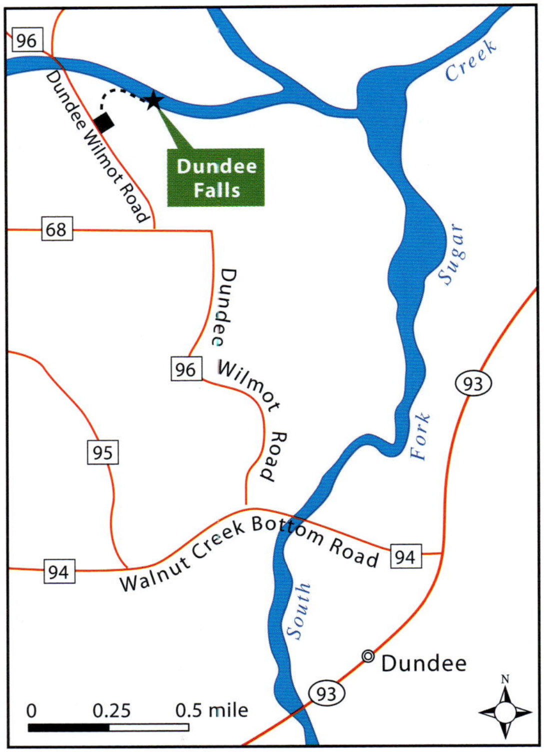

32. MASSILLON BUFFSTONE
A Rock of Fashion and Commerce

Deep in the wooded ravines of the Beach City Wildlife Area, 1.5 miles northwest of Dundee, an unnamed creek tumbles over Dundee Falls and flows into the master stream—the South Fork of Sugar Creek. Staged with a sprinkling of hemlock trees and a scattering of Massillon Sandstone exposures, this forest enclave is an ideal setting for contemplating an important ongoing industry and a once-upon-a-time style of construction.

Deposited in the Pennsylvanian Period, in an environment similar to that of today's coast of Florida, the Massillon Sandstone is distinguished by its purity: 85 percent quartz of medium-size granularity. This characteristic, along with a strong measure of cementation, makes the sandstone valuable as both refractory material and dimension sandstone, stone cut to specific sizes for use in the building industry.

The refractory industry, dependent on natural material that does not deform or change chemically at high temperatures, makes firebrick and similar heat-resistant products used to line the interiors of furnaces, kilns, and incinerators. Quartz meets the standard, and the Ohio Division of Geological Survey has sanctioned the quartz-rich Massillon Sandstone as an excellent refractory material for the iron, steel, and ceramic industries.

For almost two hundred years, brownish-red sandstone, known as brownstone, flourished as the fashion-of-choice facing for American homes and buildings. Major quarries located along the south shore of Lake Superior and within the Triassic redbeds of the Atlantic coast supplied much of the stone. Seeking relief from the high cost of shipping from these distant sources, the building industry in Ohio began, as early as 1857, to quarry Massillon Sandstone. In the late 1800s, with markets extending from New York City south to Birmingham and west to St. Louis, it was considered Ohio's most important source of brownstone. Or should it be called buffstone? Typically, rock considered brownstone weathers to a blushing rusty color due to its iron content. In contrast, Massillon Sandstone, containing minor amounts of iron, weathers to a range of subtle colors centering on a buff hue.

Though nearly obscured by weeds and brush, quarry pits are still visible in the Beach City Wildlife Area. In the late 1970s, Massillon "Buffstone" helped elevate Ohio to the position of number-one producer of dimension sandstone in the United States. The demand for this rock is now diminished, but if the need arises again, this 310-million-year-old sedimentary rock could more than adequately supply Ohio's needs for fashionable and heat-resistant stone for many years into the future.

An overhang of Massillon Sandstone at Dundee Falls.

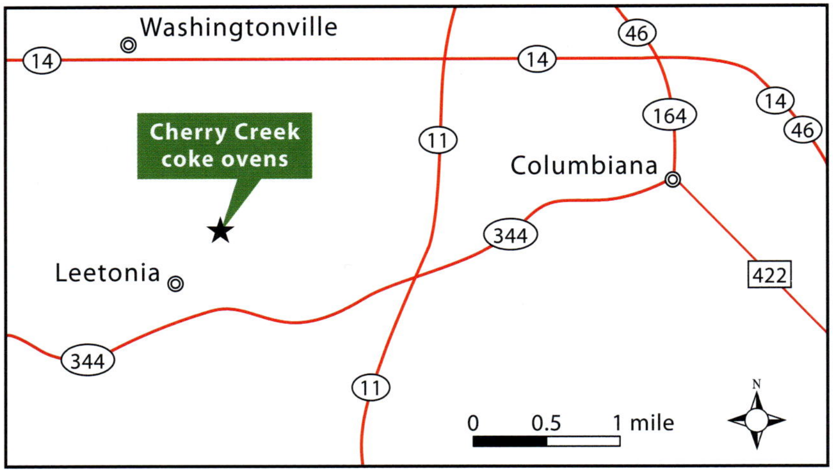

33. CHERRY VALLEY COKE
Ovens of Industry

A construction boom swept across America after the Civil War ended. The country needed iron, lots of iron, to fuel this expansion. A handful of Eastern entrepreneurs saw opportunity in this need; thus, in 1866 they formed the Leetonia Iron and Coal Company (later the Cherry Valley Iron Works) and laid out the town of Leetonia. The site they chose was ideal: the surrounding hills contained plentiful seams of bituminous coal and kidney ore, a variety of the iron-bearing mineral hematite that occurs in kidney-shaped masses—concretions or nodules. The intersection of two railroads guaranteed shipping access to the north, east, south, and west.

Their business plan called for high-volume production of coke, a hard, gray, light, porous, 90 to 95 percent pure carbon substance produced by heating bituminous coal in airtight ovens and burning out impurities, such as sulfur, tar, and moisture. The end product is well suited for use as a fuel in the manufacture of iron and steel. Initially, a 2-foot-thick vein of Pennsylvanian-age, Lower Kittanning coal, reached by a 70-foot-long incline, supplied the coke ovens. Later, a network of mine shafts intersected deeper veins, some as thick as 7 feet.

Miners obtained kidney ore by dredging gravels from nearby riverbeds. To enrich this 45-percent grade of iron ore, workers added siderite-rich shale found overlying the Kittanning coal bed. Later in the company's history, workers further enriched the ore by adding a better grade of ore shipped in from the Lake Superior Iron District.

The extensive complex of the Cherry Valley Iron Works' two hundred brick beehive ovens used to produce the coke—the largest such installation ever built in America—survive at the intersection of Butcher and Cherry Valley Roads in Leetonia. For more than sixty years the ovens cast a prosperous and healthy glow over Leetonia, producing enough coke to make 40,000 tons of miscellaneous iron products annually. Their fires were finally snuffed out in the early 1930s, a victim of the Great Depression.

Each oven is constructed of 6,800 foundry bricks set within a foundation of quarried native bedrock.

The two hundred beehive ovens in Leetonia are lined with double layers of brick. —Courtesy of Gary Phillips, Cherry Valley Beehive Coke Ovens Commission

34. KITTANNING CLAY
Potteryopolis

In 1870, when John D. Rockefeller formed the Standard Oil Company, he made Cleveland the mecca of the American petroleum industry. Three decades before that, the launching of another company in the riverfront community of East Liverpool was a lesser-known event, but of equal significance to the history of Ohio. Local businessmen opened the doors to a pottery operation—the first of many. At one time or another, more than three hundred pottery manufacturers, distributors, and decorators operated in the area, a level of industrial activity that justified East Liverpool advertising itself as the Pottery Capital of the World.

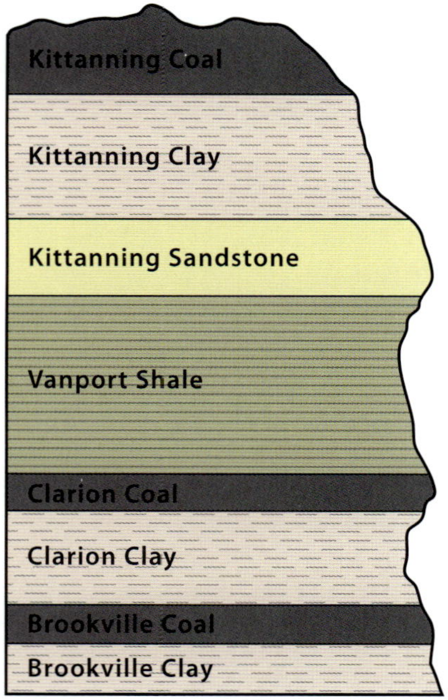

Simplified interpretation of the 1-mile-long, Ohio 7 roadcut immediately south of East Liverpool. The three clays have played an important role in the ceramics industry of eastern Ohio, but the Kittanning Clay is the most important.

East Liverpool offered two key advantages: cheap river transportation and a ready source of industrial heat. In the beginning the fuel came from hardwood forests, and then later from nearby reserves of bituminous coal and natural gas. Most important to the industry, however, was the presence of accessible deposits of clay.

Clay is the material of choice in the manufacture of firebrick, sewer pipes, terra-cotta, bricks for paving and construction, and a range of specialty chinaware such as white granite and ironstone. Defined as an unindurated sedimentary rock composed mainly of particles less than 0.00016 inch in diameter, clay's most important property is plasticity. Clay is easily shaped when wet, and once kiln fired at high temperatures it becomes hard and durable.

Economically, the most important ceramic clay found in the East Liverpool area is the Kittanning Clay of Pennsylvanian age. This clay defines the lowermost portion of a colorful, 0.5-mile-long rock exposure along Ohio 7, 5 miles south of the city limits of East Liverpool. The same clay is exposed at the top of a 1-mile-long roadcut immediately south of East Liverpool.

Exhibits of everyday ware and art ware in the Museum of Ceramics (400 East Fifth Street) depict an era when East Liverpool potteries accounted for half of the ceramics manufactured in the United States.

Kittanning Clay is exposed at the base of this roadcut 5 miles south of East Liverpool. Iron leaching from the rocks and oxidizing is the cause of the red coloration.

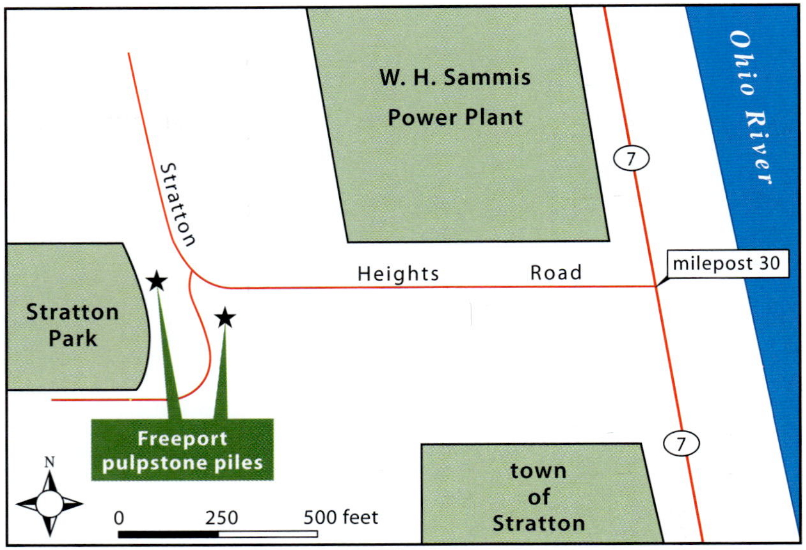

35. FREEPORT PULPSTONE
Abrasive Pulpers

About halfway between East Liverpool and Steubenville, the W. H. Sammis Power Plant and its 1,000-foot-high chimney—one of the tallest in the world—dominate the view along Ohio 7. Capable of generating enough electricity to service 1.3 million homes, this coal-fired facility has supported the local economy since 1960. Earlier, however, in the years before World War I, operations of a far different nature paid the bills and made the communities of Empire and Stratton the center of the pulpstone industry in the United States.

Pulpstones are specialized versions of millstones; they were used for grinding wood into fine fiber to make paper products. Sandstone was the rock of choice, but it had to be uniform in texture and composed of angular, not rounded, grains of quartz that were somewhat small in size. The material cementing the grains together could be either siliceous or calcareous, but it had to be firm enough to keep the finished product from crumbling under pressure. Not all sandstone fits these exacting standards, but exposures of Freeport Sandstone along the hills of eastern Jefferson County made the grade. This 70-foot-thick formation was deposited by river systems meandering throughout the area 310 million years ago, long before the nearby Ohio River came into existence.

Workers quarried square blocks of the sandstone and then shaped them into cylinders with variable thicknesses and, typically, a diameter of 54 inches. Because small amounts of interstitial water made the rock soft, they then aged the finished products for a year before putting them into service.

The pulpstone trade of Ohio, concentrated in Jefferson and Columbiana Counties, ended in the 1930s as artificial abrasives replaced their stony predecessors. Two different displays of vintage pulpstones, however, can be seen bordering the entrance to the community park in Stratton—silent sentinels of an unusual, little-remembered industry.

The diameter and thickness of the pulpstones at Stratton Park are representative of the thousands produced in Ohio between the late 1800s and 1931.

36. PUSKARICH COAL MUSEUM
The Inside Scoop on Coal

Politicians intent on expanding the boundaries of the United States carved five states and a portion of a sixth out of the landscape of the Northwest Territory during the early nineteenth century. Ohio was the first, gilded by a rich array of resources: a glacial heritage of water and fertile soil, hardwood forests that shaded the land border to border, and crude oil and natural gas fields. The greatest treasure of all was bituminous coal, a shiny, black, combustible sedimentary rock that burns with a smoky flame.

A shallow sea ebbed and flowed across the state intermittently during the Pennsylvanian Period, 323 to 299 million years ago. A series of deltas, built with sediment carried by rivers draining the rising Appalachian Mountains, created a brackish-water, shoreline environment similar to that of present-day Louisiana. Volumes of organic matter accumulated in tepid, steamy swamps and were slowly compacted into coal; a 5-foot-thick seam of coal required an accumulation of 50 feet of organic material.

Some sixty individual seams of coal have been identified in thirty-two Ohio counties, but only a few are of sufficient thickness or extent to be, or have been, economically mined. Years ago, many seams could be found exposed in the hills and hollows of southeastern Ohio. Today, however, outcrops are rare, an absence caused by surface mining, the creeping invasion of kudzu, and neighborhood growth.

Back-road travelers will find little evidence of Ohio's once-expansive coal mining industry. Artifacts of its historic significance, however, are preserved in the Harrison County History of Coal Museum in the Puskarich Public Library, at 200 East Market Street in Cadiz. Displays, memorabilia, and film presentations are available for viewing Monday through Saturday. Exhibits run the gamut, covering the geology of coal, mining methods, mining equipment, and Ida Mae Stull, America's first female coal miner. They are enclosed in walls painted jet-black, an environment intended to give the visitor the feel of a deep, dark underground mine.

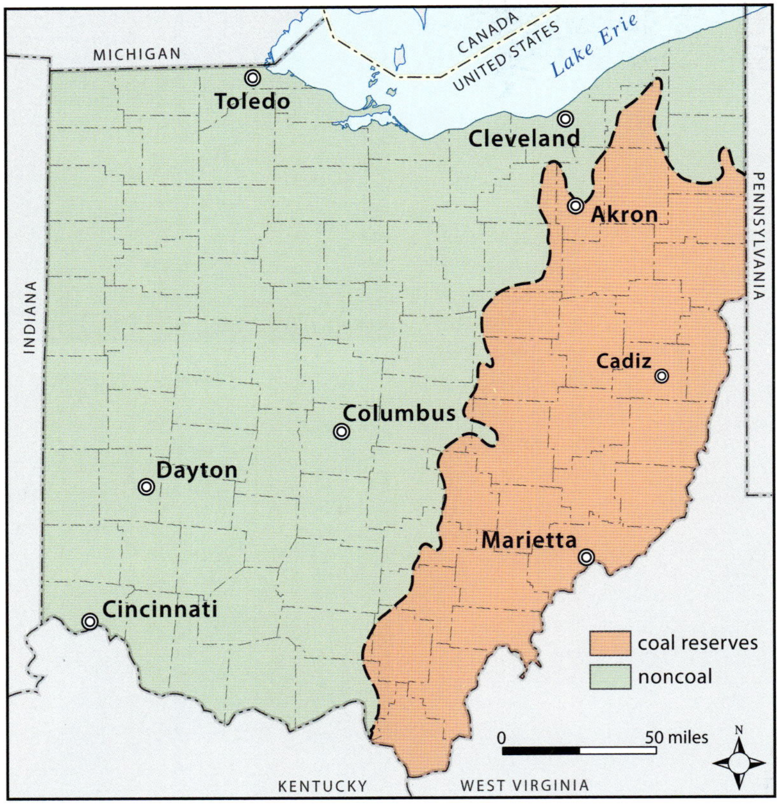

The coal region of Ohio (orange) forms the southeastern third of the state, a region defined by thick layers of Pennsylvanian- and Permian-age sedimentary rock.

An exposure of Pennsylvanian-age Upper Freeport No. 7 coal on the east side of I-77, just north of its intersection with I-70. —Courtesy of the Ohio Department of Natural Resources, Division of Geological Survey

37. MILEPOST-50 CYCLOTHEM
◆ *The Seas Roll In, the Seas Roll Out*

Sea level change is the stuff of modern-day reporting. The details are disturbing: warmer temperatures, melting glaciers, and rising sea level. To geologists, however, these trends are yesterday's news. After all, the multiple chapters of Earth history can be condensed into one simple statement: the seas advanced and the seas retreated. A roadcut along the west side of I-77 at milepost 50 provides evidence that today's changing sea level reflects many similar occurrences in the distant past.

Prior to the Pennsylvanian Period, shallow seas periodically covered much of Ohio. During Pennsylvanian time, however, sea level change was the order of the day, caused by the rise of the Appalachian Mountains and the waxing and waning of southern hemisphere glaciers. The milepost-50 exposure is ordinary in appearance: buff-colored shale and sandstone overlain by a seam of coal and topped by dark shale. In interpretation, it is a beautiful example of a cyclothem: a repetitive sequence of sedimentary rock deposited under alternating marine and nonmarine conditions.

The strata at road level represent nonmarine beds of mud and sand deposited across a floodplain by rivers draining eastern Ohio 310 million years ago. As saltwater advanced across the region, abundant plant material accumulated in coastal swamps and was later compacted into seams of coal. As the sea grew deeper, dark-hued shale, rich in the remains of marine fauna, was deposited. This black shale represents the maximum extent of the sea. Younger layers of rock, absent here but present in nearby outcrops, represent the retreating-sea portion of this cyclothem.

Before the advent of this cyclothem, similar marine invasions and retreats took place in Ohio during Pennsylvanian time, and afterward others occurred. The milepost-50 rock layers are convincing proof that sea level fluctuations have been taking place across continents for a long, long time and will, in all probability, continue to affect portions of Earth's landscapes far into the future.

The milepost-50 cyclothem tells the geologic story recorded during the first half of one of many episodes of sea level rise and fall.

38. BLACKHAND GORGE
Sandstone Shortcut

Much of the rugged topography and overall scenic beauty of the eastern Ohio countryside is influenced by massive exposures of Black Hand Sandstone, a 200-foot-thick sedimentary rock deposited during the Mississippian Period as the shallow waters of an epicontinental sea slowly retreated from the region. Crossbeds and multiple lenses of pebbles are evidence this rock was deposited in deltas, where streams and rivers entered the sea and dropped their sediment loads.

Millions of years later, prior to the onset of glaciation, rivers flowing to the west, off the developing Appalachian Mountains, drained across the Black Hand Sandstone terrain. Then, in the not-too-distant geologic past, glacial ice advanced from the north and dammed this drainage system, creating a large lake. Torrents of glacial meltwater draining from this lake eroded the bedrock, sculpting, among other landforms, the narrow east-west-trending Blackhand Gorge, which is preserved within the Blackhand Gorge State Nature Preserve.

In addition to displaying some of Ohio's most striking scenery, Blackhand Gorge is also a hub of Ohio transportation history. As long ago as 12,000 BC, Native

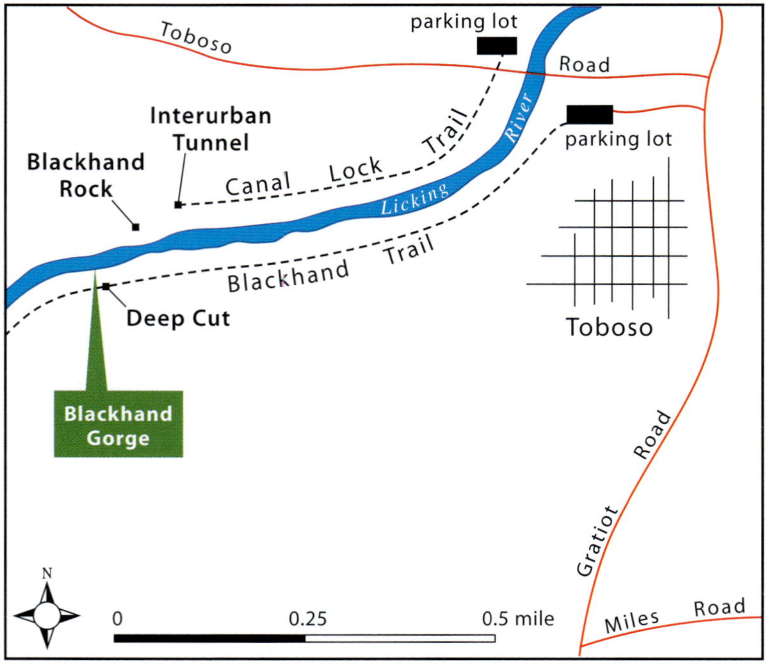

The Interurban Tunnel through Blackhand Gorge.

The smooth wall of Blackhand Rock towers over a towpath of the Ohio and Erie Canal. —Courtesy of the Ohio Department of Natural Resources, Division of Geological Survey

Americans forged a trail along the 320-foot-high walls of the gorge. In 1825, construction began on the Ohio and Erie Canal, connecting Lake Erie with the Ohio River. A short segment of the towpath of this canal is preserved behind the hand-built stone wall at Blackhand Rock. Rails of the Central Ohio Railroad were laid in 1850 through the Deep Cut, a 700-foot-long, 65-foot-high chasm hand carved through the Black Hand Sandstone. Finally, the Granville to Zanesville electric interurban rail line was tunneled 327 feet through the sandstone in 1903.

Today, asphalt and gravel trails meander beside and through these relicts of bygone years, allowing the casual visitor an opportunity to experience history that started over 300 million years ago in the Mississippian Period and continues today as the Licking River erodes even deeper into Blackhand Gorge.

Due to the massive nature of the Black Hand Sandstone, a full year was required to create the Deep Cut.

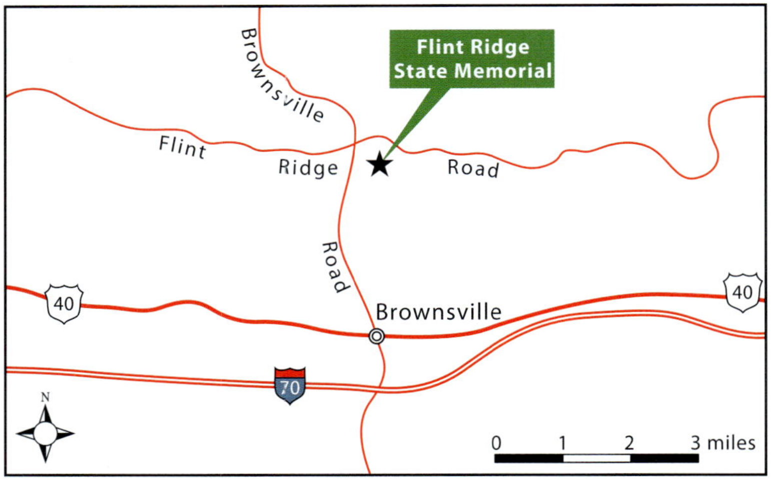

Conchoidal fractures are clearly seen in these multicolored examples of partially knapped tools. Leaf for scale.

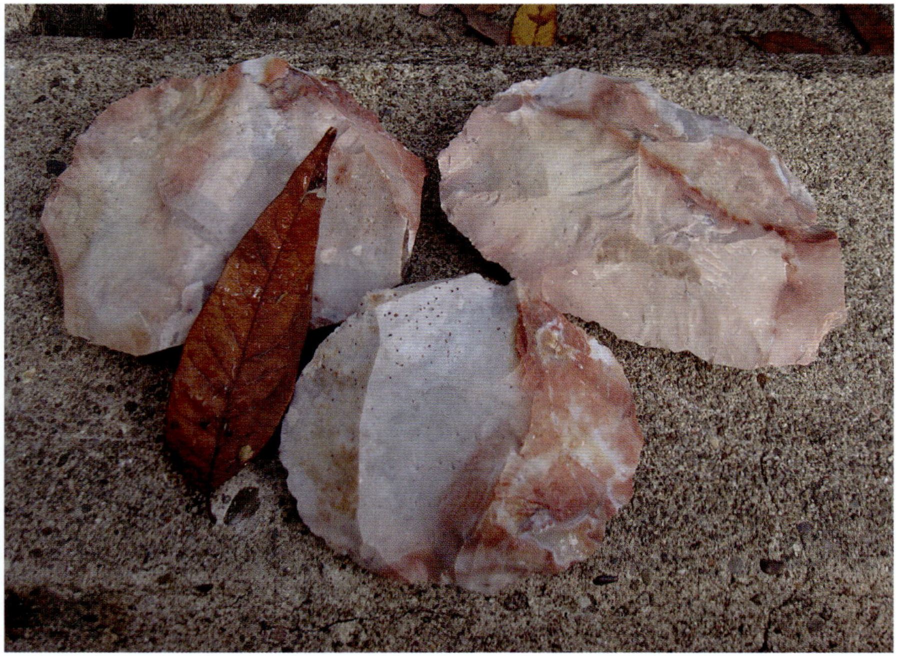

39. FLINT RIDGE
Ohio's Original Quarry

Within various circles, the term *flint* invariably creates discussion rife with controversy. To the mineralogist it is a hard version of chalcedony, defined as quartz composed of microscopically small crystals. The layperson considers *flint* a synonym for *chert*, a hard, fine-grained, quartz-rich sedimentary rock. An archeologist uses the term to describe a collection of handwrought stone artifacts. Two facts, however, are beyond dispute: flint is recognized as Ohio's official gemstone, and over 12,000 years ago Native Americans extensively surface mined it at what is now Flint Ridge State Memorial.

Collectors have found arrowheads, spear points, drills, knives, and scrapers derived from Flint Ridge along the East Coast, in Louisiana, and throughout the central Great Plains. It likely was a highly prized material because of its luxurious waxy luster, hardness, and tendency to break along a conchoidal fracture, a smooth, curving surface. Conchoidal fracturing is what allows flint to be easily fashioned into knives and scrapers with sharp edges.

Four types of flint are present in the Flint Ridge State Memorial area: massive white; bluish gray; massive with alternating brown, red, and gray laminations; and the spectacular rust-stained, porous version characterized by irregularly shaped cavities lined with minute crystals of chalcedony. Chemical analysis shows that the flint from

Flint Ridge is 96 to 99 percent pure silica, a rank that is among the highest for flint worldwide.

The discontinuous exposures of flint found across several square miles of Flint Ridge are classified as the uppermost unit of the Vanport Limestone, a formation deposited in a sea that temporarily inundated eastern Ohio during the Pennsylvanian Period. Though the jury is still out concerning the origin of silica-rich flint, two theories hold traction in the geologic world: it formed either through the process of mineral replacement in carbonate rock or by the alteration of silica-rich mud partially derived from needlelike rods that form the skeletal structure of sponges.

Throughout summer weekends, mineral enthusiasts can be found at Flint Ridge knapping flint into a variety of implements. For a nominal fee you can test your skill at this ancient art, which requires a little bit of time, and patience.

Rust-stained, porous flint with cavities lined with minute crystals of chalcedony. Penny for scale.

40. AMES MARKER BED

A Rock to Set Your Bearings By

The Interstate Highway System has been a boon to the geologic community. Old cross-country roads avoid mountainous terrain or cross it with costly switchbacks. In contrast, interstates conquer hilly territory by frontal attack, creating hundreds of roadcuts. These exposures of naked rock are pages of geologic history made available for the passing geologist to read and interpret. A good example occurs on both sides of I-77 at milepost 28.2, south of Belle Valley.

Between Akron and Marietta, roadside exposures display four rock assemblages that make up the Pennsylvanian Period in Ohio. Between milepost 120 and the intersection with I-70 are grayish sandstone and shale of the Allegheny and Pottsville Groups, their coloring and fossil content evidence of deposition in offshore marine conditions. Farther south, as far as milepost 5, the reddish coloration of the Conemaugh and Monongahela Groups indicates onshore deposition in swamps, lakes, and floodplains. The coloration is due to the presence of iron minerals. After deposition, wetland environments protected the iron in the sediment from the atmosphere; when later exposed to the atmosphere, the iron oxidized to rust-colored hues.

The crumbly ledge of tan Ames Limestone lying near the top of the milepost-28.2 roadcut identifies the approximate transition, both temporal and geographical, between older offshore strata to the north and younger onshore exposures to the south. The Ames is the product of the most extensive of several marine invasions that inundated Ohio during the Pennsylvanian Period. After it was deposited, the sea withdrew from Ohio and never returned.

The Ames Limestone is a marker bed, a distinctive, easily identified layer of rock that can be found across a wide area. It can be traced from Kentucky, through eastern Ohio, and into western Pennsylvania, northern West Virginia, and western Maryland. This widespread distribution makes it extremely valuable in the identification and mapping of rock formations of similar ages. Marker beds exist throughout the United States, but only a few are distinctive enough to be recognized when traveling at interstate speeds. The Ames lies near the top of the list.

The milepost-28.2 outcrop of Ames Limestone.

41. THORLA-MCKEE WELL
A Snake Oil Success

Edwin Drake's 1859 discovery in northwestern Pennsylvania is traditionally recognized as the first documented oil well in the United States. A review of history, however, might suggest otherwise. Flash back to 1814, when Silas Thorla and Robert McKee drilled a salt well outside the community of Caldwell. They planned to evaporate the brine the well produced, a potentially profitable venture because salt sold for $2 per barrel.

At 200 feet, the borehole struck sandstone that contained saltwater—along with crude oil and natural gas. The developers considered their venture successful and quickly developed a thriving saltworks. The volume of oil—produced at the rate of about 1 barrel per week—was a problem, however, because it had minimal value. Turning disappointment to advantage, they decided to separate the oil from the brine and sell it as "Seneca Oil," a snake oil nostrum marketed as a cure-all for rheumatism, colds, aches, and an assortment of pains.

Two years later, they drilled a second well with the same result, and the business continued to grow until it was destroyed by fire. Today, the remains of the 1816 well—a 34-inch-diameter, oil-saturated sycamore-log casing—is preserved in a small park located at the intersection of Ohio 564 and Ohio 78, east of Caldwell. Occasionally, a blurp of oil and a bubble of gas rise to the surface, evidence that minimal production is still occurring.

A claim is occasionally made that the 1816 borehole is evidence that oil was discovered in America early in the nineteenth century. Unfortunately, the well cannot be recognized as such, because Thorla and McKee intended to find only saltwater. Edwin Drake, therefore, gets the kudos for drilling the earliest oil well because his borehole was drilled with the sole intent of discovering oil.

This broken, hollow, oil-saturated, nearly 200-year-old sycamore log encases the earliest, but not the first, borehole discovery of oil in the United States.

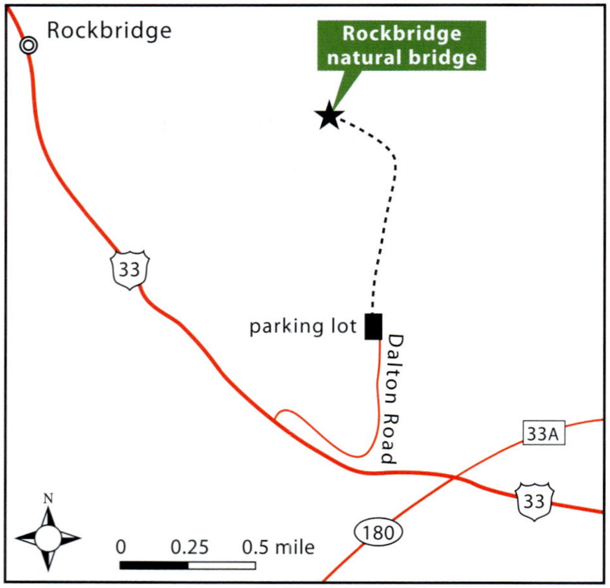

42. ROCKBRIDGE
Bridge to Nowhere

At first consideration, any natural bridge appears unnatural. How did it form and when, and how long will it last? Answers can be found at the topographic oddity forming the centerpiece of Rockbridge State Nature Preserve.

Some 350 million years ago, during the Mississippian Period, sand deposited across central Ohio by rivers was being compacted into rock—the Black Hand Sandstone. As time passed, the development of vertical fractures, the inevitable result of erosion and weathering, altered the complexion of the sandstone. Then, 2.6 million years ago, the Pleistocene ice ages began to cast a cool shadow across the countryside.

Meltwater from the glaciers flowed across the sandstone and into the fractures, forming cliffs over which developing rivers cascaded as waterfalls. The power of falling water, in harmony with percolating groundwater, slowly fashioned cavelike alcoves in the softer, lower layers of the sandstone, each recess roofed by an overhanging section of resistant rock. Rockbridge began life as one of these alcoves.

As the depth of the alcove increased to the stature of a cave and then became even deeper, portions of the cave's roof collapsed along a segment of an upstream fracture, creating a skylight. The skylight separated the main cliff face from a residual segment of the roof that remained as a lintel, which spanned a tributary of the Hocking River. Once the skylight formed, Rockbridge reached maturity.

Today, with a span of 92 feet, a maximum clearance of 40 feet, and a width that varies from 6.5 to 25 feet, Rockbridge securely holds the title of the "best known" and "longest" of all Buckeye natural bridges. It is, however, an ephemeral feature, doomed by the relentless forces of erosion and weathering. While it lasts, it gracefully spans not only a shaded sylvan brook, but also the gap between the Mississippian and Quaternary Periods of geologic time. On the other hand, considering its isolation from the hurly-burly of the modern world, maybe it is merely a bridge to nowhere.

The 40-foot drop of a Hocking River tributary highlights the steep-walled configuration of the Rockbridge skylight, evidence that one or more vertical fractures had a big influence on its formation. —Courtesy of the Ohio Department of Natural Resources, Division of Geological Survey

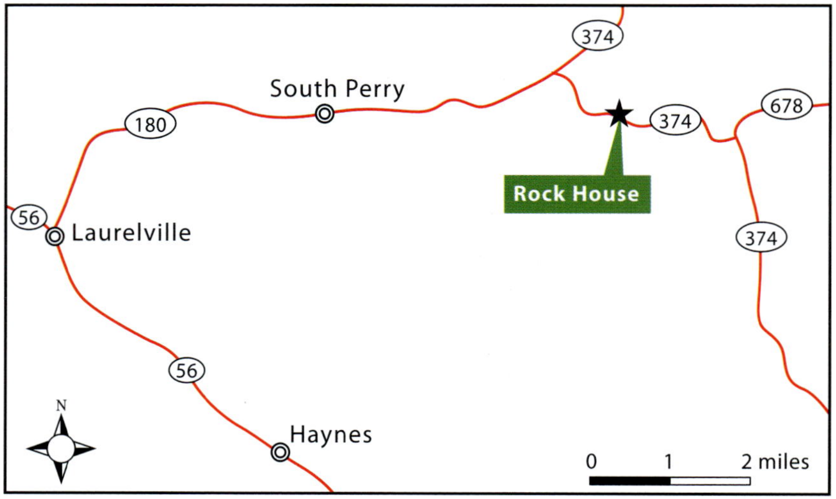

43. ROCK HOUSE
Shelter Worthy of Its Name

No doubt about it; it's an old structure in need of repair. Early residency records show that Native Americans used Rock House's recesses as energy-efficient ovens and its shallow floor troughs to collect rainwater that occasionally gushed from the leaky walls. After they moved out, a series of lowlife robbers, murderers, and horse thieves moved in, and the neighborhood went to ruin. Now it's vacant, but as a fixer-upper Rock House has potential. Consider the details: single-floor layout, thick sandstone walls, thousands of cubic feet of living space, loftlike ceiling, breezy front and rear entrances, and beautiful Gothic windows that overlook the secluded Laurel Run valley in rural Hocking County.

The foundation of Rock House is constructed of Black Hand Sandstone, named after a Native American pictograph that was destroyed in the 1820s by workers building the Ohio and Erie Canal. This massive, crossbedded sedimentary rock was deposited across central Ohio by rivers 350 million years ago, during the Mississippian Period. Over time, river erosion gave the sandstone a cliffy appearance. Its integrity was later disturbed through the formation and widening of a major fracture that encouraged a gigantic block to separate from the parent rock mass. Erosion along five equally spaced secondary joints that had developed at right angles to the master fracture created a central front entrance accentuated by adjacent sets of windows that continue to weather to pointed arches. Erosion by glacial meltwater eventually increased the interior dimensions of Rock House and gave it a high ceiling, creating an overall Gothic effect.

Rock House is even self-cleaning. It is estimated that the volume of sand necessary to fill the 120,000 cubic feet of interior space might have been removed over a 1-million-year period at the rate of 0.5 ounce per day. Wind is considered the principal housekeeping agent. In short, this early American domicile has it all: character, provenance, and location, location, and location.

Earth tones, caused by the oxidation of iron, and alcoves in the walls, caused by water erosion of the Black Hand Sandstone's secondary joints, create a bright and airy atmosphere along the 200-foot-long hallway of Rock House. —Courtesy of the Ohio Department of Natural Resources, Division of Geological Survey

44. OLD MANS CAVE
A Shadowy Tomb

It has long been an axiom among geologists that rock exposure and scenery go together like a horse and a buggy or hot dogs and the Fourth of July. Nowhere is this intimate relationship of nature more evident than along the 1-mile trail that leads to Old Mans Cave, arguably the most visited and most popular of the six scenic sections of Hocking Hills State Park. Old Mans Cave is located along Ohio 374/664, 5 miles northeast of South Bloomingville.

Here, Old Mans Creek is slowly sculpting a northeast-oriented joint into a 150-foot-deep gorge that is home to two spectacular waterfalls, a variety of bedrock oddities such as Devils Bathtub and Sphinx Head, and a 146-foot-tall hemlock, touted to be the tallest tree in Ohio.

The history of this area began millions of years before a nineteenth-century hermit made the gorge his home; supposedly he is buried beneath the overhang of Old Mans Cave. The moss-clothed Black Hand Sandstone, the dominant rock within Old Mans Gorge, was deposited by streams flowing into a shallow sea that covered eastern Ohio 350 million years ago, during the Mississippian Period. Changing environmental conditions imprinted a twofold character on this 200-foot-thick formation: the central section is less resistant to erosion than the upper and lower layers and thus weathers more quickly, forming shadow-bound recesses known as rock-shelter caves. The 75-foot-deep, 50-foot-high, and 200-foot-long Old Mans Cave is an outstanding example of this differential weathering. In contrast,

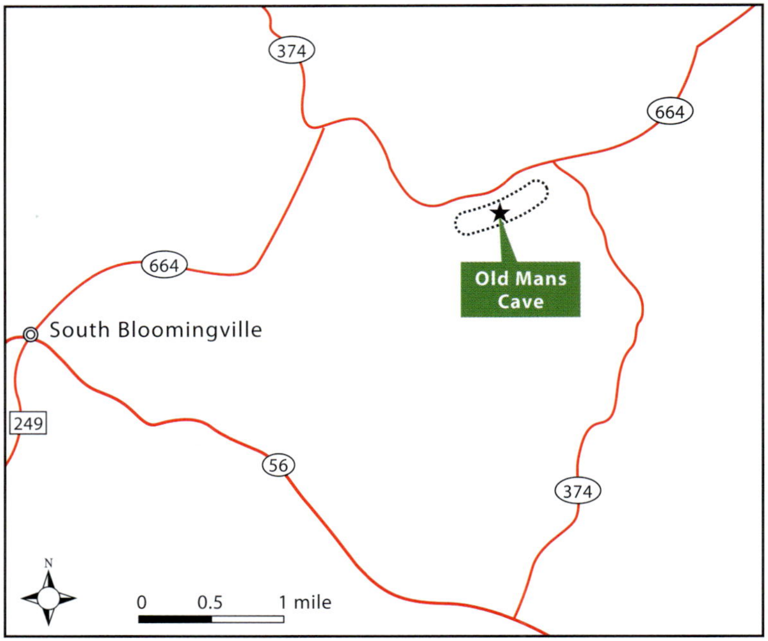

Upper Falls, 30 feet high, cascades over the upper, resistant layer of Black Hand Sandstone.

Upper Falls and Lower Falls plummet over portions of Black Hand Sandstone that are composed of well-cemented layers of nearly pure quartz.

The glaciers that invaded Ohio during the Pleistocene Epoch never quite reached this area, but their influence can be seen along the gorge in the form of towering hemlocks, yellow and black birch, and Canada yew, all biologic evidence of the frigid climate that inundated the area thousands of years ago.

Devils Bathtub is a pothole, a bowl-shaped hollow in the rocky bed of a river formed through the swirling action of coarse sediment.

The amphitheater-like recess of Old Mans Cave clearly identifies the central, erosion-prone section of Black Hand Sandstone.

45. APPALACHIAN ESCARPMENT
A Topographical Transition

The surface expression of Ohio is generally divided into three physiographic provinces, which are the basis for the organization of this field guide. The three provinces are the Lake Plain, the Till Plains, and the Appalachian Plateau. All sectors in a physiographic province are similar in geologic composition, topographic relief, and climate, and they have undergone a common history of geologic development. The contact between the Till Plains and the Appalachian Plateau is an irregular line that snakes across Ohio from northern Ashtabula County west and south to southern Adams County; it is known as the Appalachian Escarpment.

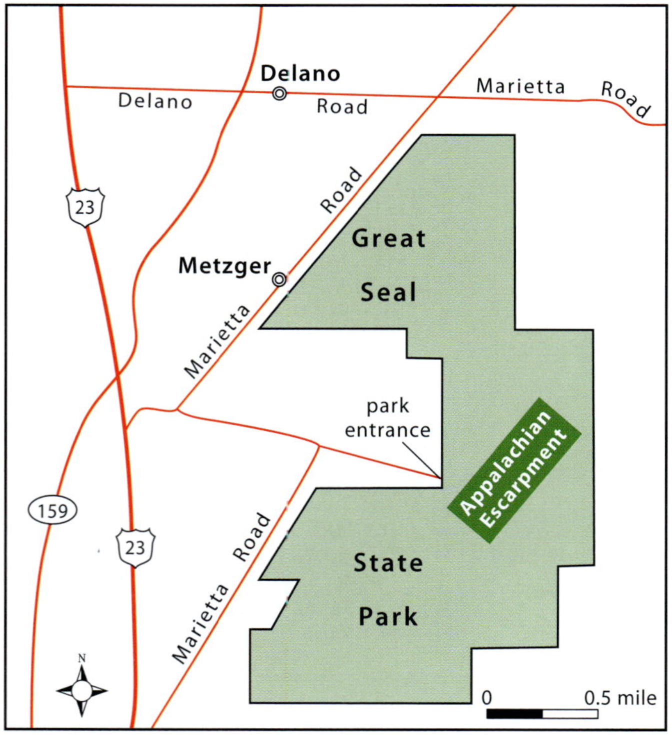

Hills and hollows, characterized by relief up to 300 feet, dominate the environs of the Appalachian Plateau. Thousands of feet of conglomerate, sandstone, and shale, interbedded with world-class reserves of bituminous coal, form its rolling, convoluted foundation of Mississippian, Pennsylvanian, and Permian bedrock. In contrast, thinner stacks of relatively flat-lying sandstone, shale, and limestone formations of Cambrian, Ordovician, Silurian, and Devonian age form the Till Plains. Oil and gas volumes here were once so vast that Ohio was designated the hydrocarbon center of the world at the dawn of the twentieth century. The thin, acidic soils of the Appalachian Plateau support forest growth, while the fertile glacial soils of the Till Plains produce abundant agricultural harvests.

There is no better place to experience this transition from semirugged topography that identifies the edge of the Appalachian Mountains to flatland country that fronts the Great Plains than along the trails of Great Seal State Park, 5 miles northeast of Chillicothe. The hills, cultivated fields, and overview of the Scioto River valley seen here are the very same vista portrayed on the Great Seal of the State of Ohio.

The Appalachian Escarpment is identified by the sharp change from flatland (foreground) to the foothills of the Appalachian Plateau (background). —Courtesy of Nyttend, Wikimedia Commons

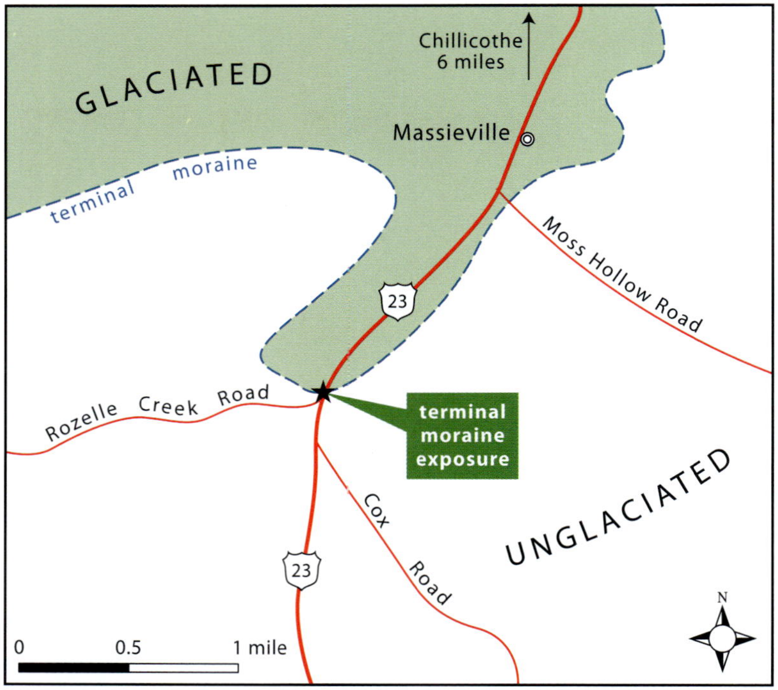

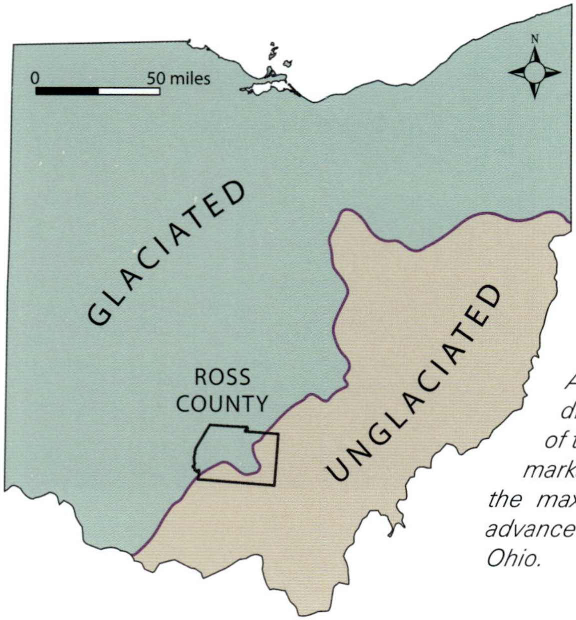

A bulge in the distribution pattern of till in Ross County marks the location of the maximum southeast advance of glacial ice in Ohio.

46. MASSIEVILLE TERMINAL MORAINE

An Ice Sheet Extremity

On a long-ago morning, the rising sun spread a warm glow over the Ohio landscape, but the effect was different. The sun failed to warm the land as much as it had the day before because an ever-growing sheet of ice was flowing outward from the region of today's Hudson Bay. As the ice sheet's footprint expanded, it altered drainage patterns and bulldozed the countryside down to nearly pool-table flatness. Worldwide sea levels fluctuated, dropping as much as 300 feet. This was the time of the great ice ages of the Pleistocene Epoch, a period of global cooling that began 2.6 million years ago and ended in Ohio about 14,000 years ago.

Four major advances of ice scraped their way to and fro across as much as two-thirds of the Buckeye State during the Pleistocene Epoch. Some 17,000 years ago higher temperatures effectively stopped further ice movement. Under a warming climate, the degree of glacial melting had become equal to ice formation. In essence, the ice stalled in place, depositing a terminal moraine—a conspicuous ridge of rock debris—along its margin before melting back from the region. *Terminal* denotes that the moraine marks the farthest advance of the glacier. This moraine is well exposed along US 23, 1.5 miles south of Massieville in the foothills of the Appalachian Mountains.

Southeast of this moraine, the ancient profile of Ohio remains in virginal condition, having remained unviolated by glacial erosion. To the north and west, thick layers of mineral-rich rock debris, deposited by ice, blanket the land. The moraine closely coincides with the transition from mountain terrain to flatland and identifies the edge of the great Appalachian forest and the beginning of the productive corn and wheat farmland known as the Bread Basket of America.

A piece of Canadian granite (left) from the terminal moraine contrasts sharply with bedrock fragments of Mississippians-age sandstone found underlying and adjacent to the moraine. Quarter for scale.

A growth of mature trees (right) behind the intersection of US 23 and Rozelle Creek Road, south of Massieville, covers the leading slope of the terminal moraine.

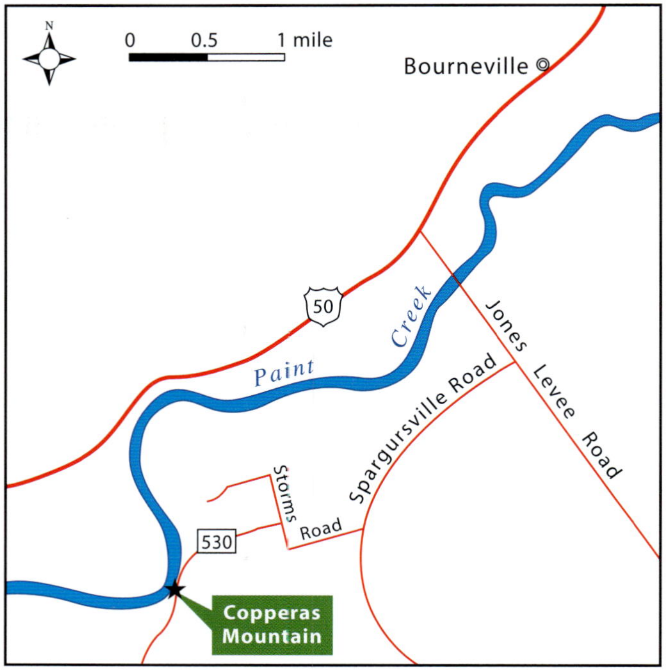

47. COPPERAS MOUNTAIN
Mineral Oddities of the First Order

Sheltered along a 350-foot-high cliff 3.5 miles—as the crow flies—southwest of Bourneville is an outcrop of Devonian-age Ohio Shale that some tout as the most extraordinary exposure of sedimentary rock in the entire state. While such a claim is open to dispute, this site at Copperas Mountain does contain a degree of dash and flash in the form of two unusual phenomena associated with minerals: concretions and efflorescence. Two prominent examples of the former, several feet in diameter, occur in the northern portion of the exposure, while the latter is on sporadic display across the full extent of the rock face.

Nearly spherical in shape, the concretions stand out because of the layers of shale that wrap around them like well-fitting sheets of insulation. The origin of these sedimentary rock structures is still debated, but most geologists agree they form from mineral matter that precipitates, from groundwater, around a small nucleus, such as a bone or shell fragment. This process takes place after the enclosing sediment has been deposited and before it is compacted to rock. Typically, the cores of these concretions are filled with coarse-grained veins of calcite and quartz.

Efflorescence, an uncommon occurrence, paints this brownish-black and greenish-gray shale with a panorama of tints ranging from snow-white to yellow to blue-green. As silt-sized grains of copperas—an iron-bearing mineral that gives the mountain its name—begin to weather, the splotches of multihued color appear. They are fluffy blooms of wispy powder left behind when water evaporates.

When the Ohio Shale is heated in the laboratory, it yields a volume of crude oil averaging 3.6 gallons per ton of rock—the darker the color, the higher the carbon content and the greater the yield. The Copperas Mountain site has it all: scientific curiosity, cosmetic coating, and economic potential.

Mineral efflorescence is similar in chemistry to the white, chalky substance seen on concrete and brick walls in damp, musty environments.

Copperas Mountain concretions have a hard, resistant dolomite exterior and a soft, fractured core that weathers easily, leaving a hollow interior. —Courtesy of the Ohio Department of Natural Resources, Division of Geological Survey

48. BUCKEYE FURNACE

Pig Iron Is Born

In 1818 Native American paths formed the basis of the primitive transportation system of Ohio. Life was difficult, and the search was on for a source of that one commodity every farmer, merchant, and homemaker needed—iron. Then, eureka! The Hanging Rock Iron Region was discovered, stretching across much of southeast Ohio and into Kentucky and West Virginia. The region became the focus of a charcoal blast-furnace industry that, for nearly a century, played a major role in making the Buckeye State an industrial giant.

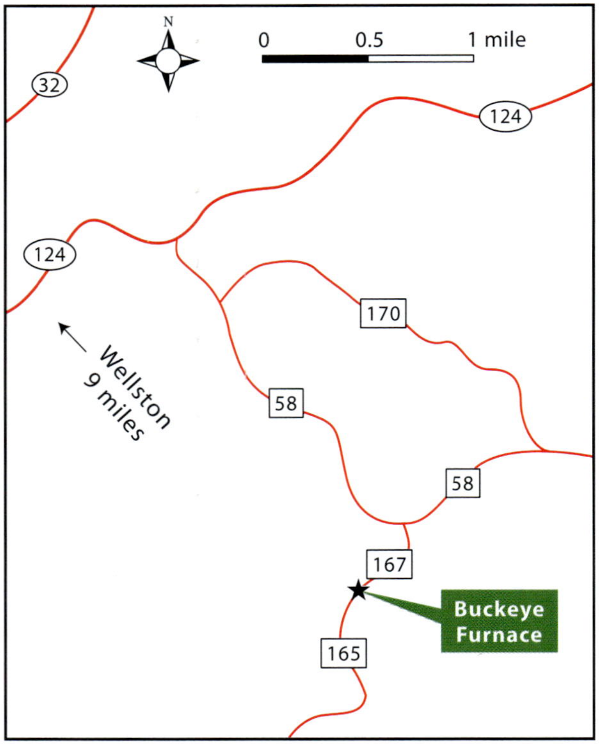

Everything the industry needed could be found nearby: deposits of ore, accessible exposures of limestone, acres of virgin timber, and even easily quarried sandstone to build the massive furnace stacks. By the 1870s, forty-six furnaces were operating in Ohio, producing more than 100,000 tons of iron annually. Today, the authentically reconstructed Buckeye Furnace, located in the sylvan woods of Jackson County, presents a detailed and colorful history of the time.

Telltale benches near this geologic site identify locales where workers scraped foot-thick layers of Little Red Block, Guinea Fowl, and Yellow Kidney ore, averaging 30 to 40 percent iron, off the hillsides. The ore layers conveniently lay on top of the Pennsylvanian-age Vanport Limestone, which was also mined for use as flux, a calcium-rich mineral or rock added to the flames to enhance the quality of the final product. On a typical day, Buckeye Furnace produced 12 tons of pig iron using 31 tons of ore, 1.7 tons of limestone, and 1,650 bushels of charcoal.

During Pennsylvanian time, streams flowing off the flanks of the Appalachian Mountains deposited iron ore in the wetlands of the Hanging Rock Iron Region. The ore, limestone, and sandstone are still there, but new technologies forced the last furnace to close in 1916. The fires have long been extinguished, but the furnaces are not forgotten. Buckeye Furnace, open seasonally, is reminiscent of that time when America was young and pig iron was king.

The reconstructed Buckeye Furnace. The building on the left contained steam pumps that blasted air into the smelting zone of the furnace, the stone structure in the center. The front building protected the furnace from the elements and was where molten iron was poured into sand molds to form pig iron. —Courtesy of the Ohio Department of Natural Resources, Division of Geological Survey

49. SCIOTO SALINE
A Frontier Necessity

In many places in the world the mineral halite is as scarce and precious as it is common elsewhere. This benign combination of explosive metal (sodium) and poisonous gas (chlorine), commonly known as salt, is essential to health and is a vital ingredient in the production of glass, the tanning of leather, and the preservation of an extensive list of food products.

As the wave of late-eighteenth-century colonization swept across Ohio, halite took on great importance because naturally occurring exposures were scarce. Pioneer families followed the habits of Native American tribes and searched for licks, which are saltwater springs. Explorers found one of the most important of these ancient licks, the Scioto Saline, along the banks of Salt Lick Creek in Jackson County.

For a period of two decades (1796 to 1815) the Scioto Saline was the centerpiece of the most important mineral industry in Ohio. Salt had been obtained there for at least 8,000 years, as evidenced by the age dating of bones of extinct animals. When the site of the earliest colonial production was absorbed by urban construction, activities were moved to nearby Boone Rocks, a 57-foot-high cliff of Pennsylvanian-age Sharon Sandstone that lies in a shaded glen on the northwest edge of Jackson.

Pioneer settlers collected brine that slowly seeped into pits they dug by hand into the Sharon Sandstone when it was exposed in the streambed during low-water stages of Salt Lick Creek. Some geologists suggest the brine is derived from a domelike structure created by the upward movement of deep-seated salt beds. They boiled the brine in 12- to 15-gallon kettles on a 24/7 basis. Gathered in baskets, the salt precipitate sold for $2.50 per bushel, the equivalent of $50 today. After more concentrated brines were discovered around 1810 along the Kanawha River in what was then the state of Virginia, activity in the Jackson area declined. Today, little evidence of this frontier industry remains. Nonetheless, in the heyday of the Scioto Saline, individuals who daily labored there were most certainly worth their salt in more ways than one.

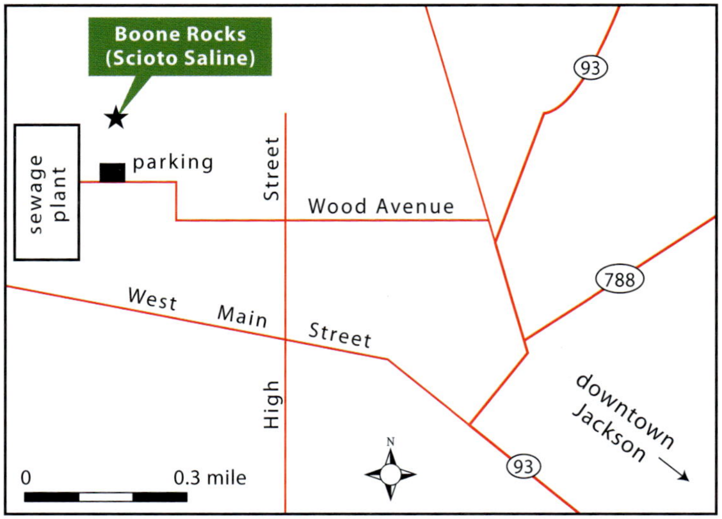

Boone Rocks, the exposure of Sharon Sandstone that became the site of salt production after the earlier locale was enveloped by urban development. —Courtesy of the Ohio Department of Natural Resources, Division of Geological Survey

50. TEAYS RIVER VALLEY
Phantom River of Ohio

The 981-mile-long, southwesterly flowing Ohio River drains approximately three-quarters of the Buckeye State. Such was not always the case. Prior to the onset of glaciation 2.6 million years ago, the 1,100-mile-long Teays River sculpted the majority of Ohio's landscape.

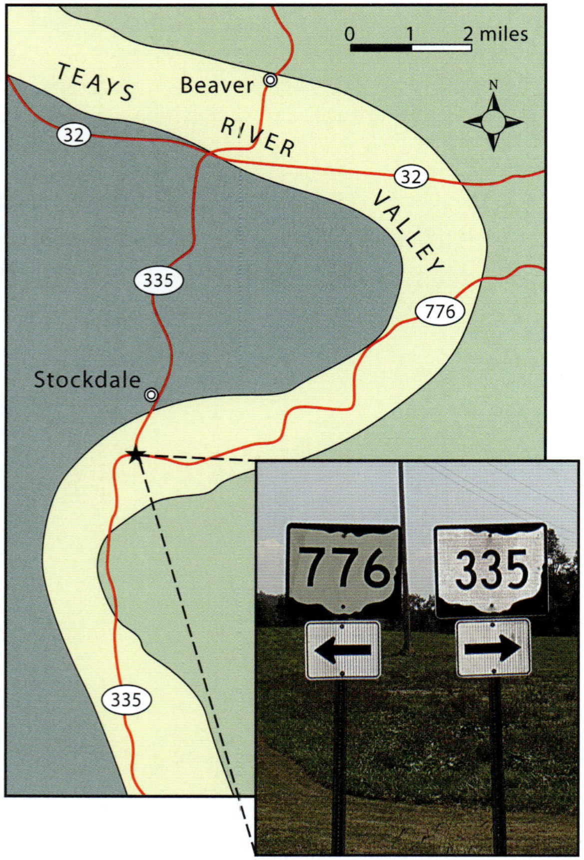

The birthday of this river of intrigue took place some 5 million years ago, during the Neogene Period, when the plains of Ohio and the lofty peaks of the nearby Appalachian Mountains dominated the eastern landscape of North America. From headwaters located in western North Carolina, near the mountain town of Blowing Rock, the Teays River flowed north across Virginia and West Virginia and meandered northwesterly across Ohio. Crossing Indiana and Illinois, it finally joined an ancient embayment of the Gulf of Mexico now occupied by the Mississippi River. In its heyday the Teays River valley was as much as 2 miles wide and up to 500 feet deep.

After Pleistocene-age ice reached its farthest advance in southeastern Ohio, a massive dam of glacial debris blocked the course of the Teays. The result was a 7,000-square-mile lake—nearly the size of Lake Erie—that inundated southern Ohio and parts of West Virginia and Kentucky. When this lake (called Lake Tight) overflowed, the present-day Ohio River drainage system began to form and the Teays River system died.

Today, much of the Teays River valley lies beneath glacial sediment. Between Wheelersburg and Waverly, however, a question-mark-shaped segment is visible at the surface. At the intersection of Ohio 776 and Ohio 335, in northern Scioto County, the forested banks of the Teays River valley are easily observed to the east and west. There, on a quiet day and with a little imagination, the rumble of torrents of turbid water can again be heard—the sounds of a colossal, phantom river that once dominated and drained the landscape of eastern North America.

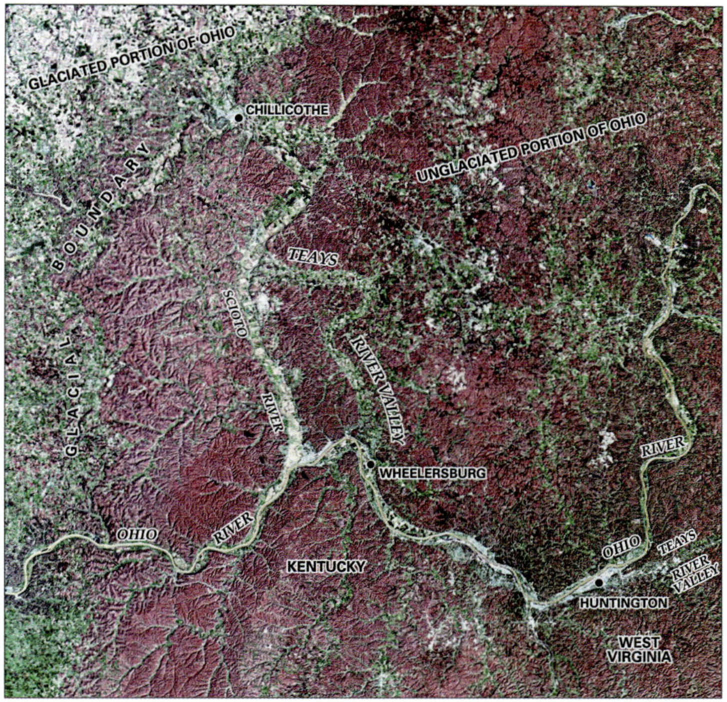

An abandoned segment of the Teays River valley between Wheelersburg and Waverly is clearly illustrated as a green valley on this Landsat 7 satellite image. The modern Ohio River follows the ancient Teays River valley from Huntington, West Virginia, to Wheelersburg.
—Courtesy of the Ohio Department of Natural Resources, Division of Geological Survey

Looking east from the intersection of Ohio 335 and Ohio 776, the hills on the horizon mark the southeastern bank of the Teays River valley.

GLOSSARY

Acadian Orogeny. The second of three phases of mountain building that were instrumental in the development of the Appalachian Mountains during the Paleozoic Era.

algae. Primitive aquatic plants that lack true stems, roots, and leaves.

Alleghanian Orogeny. The third of three phases of mountain building that were instrumental in the development of the Appalachian Mountains during the Paleozoic Era.

amniotic egg. An egg produced by reptiles, birds, and egg-laying mammals in which the embryo develops inside a membranous sac.

anhydrite. A mineral composed of anhydrous (no water) calcium sulfate.

basement. The undifferentiated complex of typically crystalline rocks underlying the sedimentary deposits in a region.

bed. The smallest sedimentary unit that is distinguishable from those above and below it based on sediment type, sediment size, and/or sediment color.

bedding plane. The surface of separation between any two distinct beds of sedimentary rock.

bedrock. A general term for rock, usually solid, that underlies soil or other unconsolidated, superficial material.

biomass. The amount of living material in a particular area, stated in terms of the weight or volume of organisms per unit area.

bituminous coal. A dark-brown to black variety of coal that burns with a smoky flame and is the most abundant rank of coal in the United States.

bog. Waterlogged ground containing acidic, decaying vegetation that may develop into a low grade of coal.

brachiopod. A marine invertebrate with bivalve shells enclosing arm structures used to sweep food into the mouth. Also called a *lampshell.*

bryozoan. A small aquatic invertebrate that feeds with a crown of hollow tentacles and has a calcareous skeleton. It forms colonies. Also called a *moss animal*.

calcareous. Consisting of calcium carbonate.

calyx. The plated, skeletal structure surrounding the body of a crinoid.

carbonate. A mineral, such as calcite or aragonite, containing a carbonate ion. Also a sedimentary rock, such as limestone, chalk, or dolostone, composed of appreciable amounts of carbonate minerals.

cephalopod. A marine organism with an internal shell, such as an octopus or nautilus.

charcoal. An impure carbon residue created by burning organic material in the absence of air.

clay. A sedimentary particle less than 0.00016 inch (0.004 mm) in diameter.

conglomerate. A sedimentary rock composed of rounded pebbles, cobbles, and boulders held together by a fine-grained matrix.

coral. A general name for a large group of bottom-dwelling invertebrate organisms common in warm seas. They typically live in colonies; some species form reefs.

core. The central zone of the Earth's interior, below a depth of 1,800 miles. It is divided into an inner, solid subcore and an outer, fluid subcore.

crinoid. A marine animal characterized by a cup-shaped body, feathery arms, and a clawlike stem attached to the seafloor. Also known as *sea lilies* or *feather stars*.

crossbed. A sedimentary rock layer inclined at an angle to the main planes of stratification.

crust. Earth's outermost shell, which varies in thickness between 6 and 25 miles.

crystal. The solid body of a chemical compound having a regularly repeating atomic arrangement that may be outwardly expressed by plane faces.

dissolution. The dissolving of rock through chemical alteration.

dolomite. A common rock-forming carbonate mineral composed of calcium and magnesium.

dolostone. A sedimentary rock composed principally of the mineral dolomite.

drift. A general term for all rock material transported and deposited by a glacier.

dune. A usually asymmetrical ridge of sand that has been transported and deposited by wind.

erratic. A relatively large rock fragment carried by glacial ice, deposited some distance from the outcrop from which it was derived, and generally resting on bedrock of a different nature.

exoskeleton. A supportive and protective framework that lies outside the body tissues of an animal, forming an external armor.

fault. A break in rock layers along which there has been movement on one side relative to rock on the adjacent side.

fauna. The entire animal population, living or fossil, of a given area, environment, or time span.

firebrick. A heat-resistant brick used for lining furnaces, chimneys, and fireplaces.

flora. The entire plant population, living or fossil, of a given area, environment, or time span.

formation. The basic cartographic rock unit in geologic mapping. A formation has easily recognizable boundaries based on its rock characteristics that can be traced along an outcrop in the field and in the subsurface.

fossil. The remains or evidence of preexisting life.

gastropod. A coiled-shell organism, such as a snail, slug, or limpet.

genus. A unique group of organisms with one common characteristic or several common characteristics.

geode. A spherical body of rock with a cavity that is partly filled with a lining of inward-projecting crystals deposited from solution.

glacier. A mass of ice that moves outward in all directions due to gravity and the stress of its own weight.

granite. A coarse-grained (visible mineral crystals), light-colored igneous rock that formed from magma below the surface and is rich in quartz and potassium feldspar.

Grenville Mountains. The mountain range, formed by the collision of several microcontinents late in the Proterozoic Eon, that is the geologic core of eastern North America.

groundwater. A general term for all subsurface water, as distinct from surface water.

gypsum. A sedimentary rock composed of calcium sulfate that precipitates from evaporating water.

halite. A common, usually colorless mineral composed of sodium and chlorine and having a distinctive salty taste.

hematite. A common mineral that is the principal ore of iron.

hollow. A synonym, used especially in the southeastern United States, for a narrow, sheltered valley.

host rock. A body of rock serving as a host for other rocks or for mineral deposits.

ice sheet. A glacier of considerable thickness covering a large part of a continent.

igneous. A class of rock formed by the crystallization of magma.

karst. Topography formed on and within a carbonate rock foundation and characterized by caves, sinkholes, and other dissolution features.

kettle. A steep-sided, bowl-shaped depression in glacial-drift deposits formed by the melting of a detached block of stagnant ice left behind by a retreating glacier. Often contains a lake or swamp,

lens. A rock deposit that is thick in the middle and tapers to nothing at its edges.

limestone. A sedimentary rock composed chiefly of calcium carbonate. Typically forms in lakes and warm, shallow seas.

limonite. A general term for a mineral group of brown, amorphous, naturally occurring hydrous ferric oxides.

lithosphere. The solid, outer portion of Earth, as compared with the atmosphere and the hydrosphere.

magma. Molten or partially molten rock found in the interior of Earth. Called *lava* when it erupts onto the surface.

magnetic analysis. The study of a body of rock based on a review of its magnetic properties.

mantle. The zone of the Earth that lies below the crust and above the core.

massive. Said of rock with homogenous texture and lacking layering, fractures, or other discontinuities.

meltwater. Water derived from melting snow or glacial ice.

metamorphic. A class of rock that has been changed in appearance due to exposure to excessively high temperatures and pressures.

moraine. A ridge of drift that accumulated along the front and sides of a glacier.

mud crack. An irregular fracture with a polygonal pattern formed by the shrinkage of clay, silt, or mud.

ore. Naturally occurring material from which a mineral or minerals of economic value can be extracted.

orogeny. The process of the formation of mountains.

outcrop. The part of a formation or geologic structure that is exposed at Earth's surface.

outwash plain. A broad, gently sloping sheet of sediment deposited by meltwater flowing in front of a glacier.

oxidation. The combination of a substance with oxygen.

Pangaea. The supercontinent that formed late in the Paleozoic Era through the merger of the protocontinents Gondwanaland and Laurasia.

pebble. A generally rounded stone between 0.16 and 2.5 inches (4 and 64 mm) in diameter.

pelecypod. A shelled marine or freshwater organism, such as a clam or an oyster.

phylum. The primary taxonomic division of the animal kingdom.

physiographic province. A region with a unified geomorphic history, wherein all parts are similar in geologic structure and climate.

pig iron. Crude iron that is the direct product of a blast furnace and is refined to produce steel and cast iron.

plate. A segment of Earth's lithosphere that moves horizontally to, and independently of, other plates and interacts with them along zones of seismic activity.

plate tectonics. The theory that most large features of Earth form through the relative movement and interaction of the rigid plates composing the lithosphere.

protocontinent. A landmass that might later be enlarged into a major continent.

refractory. A material that does not deform or change chemically when subjected to very high temperatures.

ripple mark. An undulatory rock surface composed of small-scale ridges and hollows that formed as wind or flowing water moved sediment around before it hardened into rock.

Rodinia. The supercontinent that formed toward the end of the Precambrian Era when the protocontinents of South America, eastern North America, and western Africa merged.

rugose coral. A type of extinct coral that had wrinkled or ridged surfaces.

sand. A sedimentary particle ranging in size between 0.0025 and 0.08 inch (0.06 and 2 mm) in diameter.

sandstone. A sedimentary rock composed of rounded, sand-sized sediment that is held together by naturally occurring cement.

scale tree. A fossil tree of Carboniferous age with an exterior marked by diamond-shaped scars left after leaves fell off the tree.

seam. A particular bed or vein in a series of beds, such as a coal seam.

sediment. Unconsolidated solid particles that originate from the weathering and erosion of rock.

sedimentary. A class of rock formed by the deposition and cementation of sediment.

seismic. Pertaining to an earthquake or Earth vibration, including those that are artificially produced.

seismicity. The phenomenon of Earth movements.

shale. A sedimentary rock composed of clay- and silt-sized sediment. Has a tendency to break along thin, parallel planes.

siderite. A mineral that is a valuable ore of iron.

silt. A sedimentary particle ranging in size between 0.00016 and 0.0025 inch (0.004 and 0.06 mm) in diameter.

siltstone. A sedimentary rock composed of silt-sized sediment with a texture that is intermediate between sandstone and shale.

sinkhole. A circular topographic depression in an area of karst.

sorting. The distribution of grain size in sediment or sedimentary rock. *Well sorted* refers to sediment such as beach sand that has been sorted by wind or water. *Poorly sorted* refers to sediment such as glacial till that has not been sorted by wind or water.

speleothem. Any of the various types of mineral deposits that form in caves.

stalactite. A speleothem that hangs from the ceiling of a cave and develops as minerals precipitate from dripping water.

stalagmite. A speleothem that grows upward from the floor of a cave and develops as minerals precipitate from water that drips onto the floor.

strata. Tabular layers of sedimentary rock separated from each other by a distinctive bedding plane.

supercontinent. A landmass formed periodically by the amalgamation of all the existing continents.

suture. A fault that marks the boundary between two plates that were once widely separated.

Taconic Orogeny. The first of three phases of mountain building that were instrumental in the development of the Appalachian Mountains during the Paleozoic Era.

tectonics. Pertaining to the forces involved in the development of the broad architecture of Earth's crust, such as its ocean basins, mountains, folds, and faults.

terminal moraine. The moraine that marks the farthest advance, or maximum extent, of a glacier.

terrain. A region of the Earth that is considered a physical feature, such as the Appalachian Plateau.

till. Unsorted and unconsolidated drift deposited beneath a glacier without being reworked by meltwater.

Tippecanoe Sea. An arm of an extensive sea that covered much of what is today North America during the Ordovician and Silurian Periods.

trilobite. An extinct marine arthropod characterized by a three-lobed exoskeleton.

Wisconsinan stage. The fourth glacial stage of the Pleistocene Epoch in North America.

REFERENCES

TILL PLAINS

Atwood, W. W. 1940. *The Physiographic Provinces of North America*. Ginn and Company.

Birdsall, S. S., et al. 2008. *Regional Landscapes of the United States and Canada*. John Wiley and Sons, Inc.

De Blij, H. J., and P. O. Muller. 2010. *Geography: Realms, Regions, and Concepts*. John Wiley and Sons, Inc.

1. Seneca Caverns, Seneca County

Page, T. 1968. *A Report on the Amphipods of Seneca Caverns*. Kent State University.

Ruedisili, L. C., et al. 1990. "Geology of Seneca Caverns, Seneca County, Ohio." *Ohio Journal of Science* 90 (4): 106–11.

White, G. W. 1926. "The Limestone Caves and Caverns of Ohio." *Ohio Journal of Science* 26 (2): 73–116.

2. Indian Trail Caverns, Wyandot County

Hansen, M. C. 1992. "Indian Trail Caverns—A Window on Ohio's Pleistocene Bestiary." *Ohio Geology Newsletter* spring: 1–3.

Hobbs, H. H., and K. M. Kissell. 2010. "A Description of Sheriden Cave, Wyandot County, Ohio." *Pholeos* 29: 17–21.

Tankersley, K. B. 1997. "Sheriden: A Clovis Cave Site in Eastern North America." *Geoarchaeology* 12 (6): 713–24.

3. Oesterlen Gas Discovery, Hancock County

Orton, E. 1888. "The Trenton Limestone as a Source of Petroleum and Inflammable Gas in Ohio." In *Ohio Division of Geological Survey*, vol. 6, 101–310.

Wickstrom, L. H., et al. 1992. *Stratigraphy, Structure, and Production History of the Trenton Limestone (Ordovician) and Adjacent Strata in Northwestern Ohio.* Report of Investigations 143. Ohio Department of Natural Resources, Division of Geological Survey.

4. First Offshore Oil Well, Auglaize County

Dickas, A. B. 2010. "Shelby County and the Offshore Drilling Industry." *Historical Highlights*, Shelby County (Ohio) Historical Society, July: 3–5.

Sneed, J. L. 2005. "The Lost History of Ohio's Grand Reservoir Oil Boom." *Oil-Industry History* 6 (1): 49–53.

5. Bellefontaine Outlier, Logan County

Hansen, M. C. 1991. "Campbell Hill—Ohio's Summit." *Ohio Geology Newsletter* winter: 1–5.

Smith, G. H. 1935. "The Relative Relief of Ohio." *Geographical Review* 25 (2): 272–84.

Wickstrom, L. H. 1990. "A New Look at Trenton (Ordovician) Structure in Northwestern Ohio." *Northeastern Geology* 12 (3): 103–13.

6. Big Rock, Shelby County

Dickas, A. B. 2007. "The Rest of the Big Rock Story." *Historical Highlights*, Shelby County (Ohio) Historical Society, December: 3–6.

———. 2012. *101 American Geo-Sites You've Gotta See.* Mountain Press Publishing Company, 146–47.

Hansen, M. C. 1984. "Glacial Erratics, or 'What's the Biggest Rock in Ohio?'" *Ohio Geology Newsletter* winter: 1–5.

Hussey, J. 1878. "Geology of Shelby County." In *Report of the Geological Survey of Ohio*, vol. 3, 448–67, Nevins and Myers State Printers.

7. Lockington Locks, Shelby County

Morthorst, M. E., 1997. *The Miami and Erie Canal and the Sidney Feeder Canal in Miami County, Ohio, and Shelby County, Ohio*. Canal Society of Ohio.

Stephan, N. 2006. *Low Bridge: A Saga of Ohio's Miami-Erie Canal*. PublishAmerica.

Triplett, B. 2008. *Canals of Ohio: A History and Tour Guide*. CreateSpace Independent Publishing Platform.

8. Ohio Caverns, Champaign County

Ohio Caverns: Ohio's Outstanding Natural Wonder. Distributed by Ohio Caverns, West Liberty, Ohio.

White, G. W. 1926. "The Limestone Caves and Caverns of Ohio." *Ohio Journal of Science* 26 (2): 73–116.

9. Huron Boulders, Delaware County

Barth, V. D. 1975. "Formation of Concretions Occurring in the Ohio Shales along the Olentangy River." *Ohio Journal of Science* 75 (3): 162–63.

Clifton, H. E. 1957. "The Carbonate Concretions of the Ohio Shale." *Ohio Journal of Science* 57 (2): 114–24.

Hansen, M. C. 1994. "Concretions: The 'Ludus Helmontii' of the Ohio Shale." *Ohio Geology* fall: 1–6.

10. Ripple Rocks, Franklin County

Hansen, M. C. 1999. "The Geology of Ohio—The Devonian." *Ohio Geology* 1: 1, 3–7.

Hoover, K. V. 1960. *Devonian-Mississippian Shale Sequence in Ohio*. Information Circular 27. Ohio Department of Natural Resources, Division of Geological Survey.

11. Charleston Falls, Miami County

Hansen, M. C. 1997. "The Geology of Ohio—The Ordovician." *Ohio Geology* fall: 1–6.

Schumacher, G. A. 2008. "How Oceans, Rivers, and Glaciers Created the Geology of the Dayton Region." *Ohio Geology* 1: 1, 3–5.

12. Oakes Quarry, Greene County

Ausich, W. I. 1986. "Early Silurian Inadunate Crinoids (Brassfield Formation, Ohio)." *Journal of Paleontology* 60 (3): 719–35.

———. 1987. "Brassfield *Compsocrinina* (Lower Silurian Crinoids) from Ohio." *Journal of Paleontology* 61 (3): 552–62.

Wolfe, M. E. 2008 "All Aboard for a Travel Back in Time at Oakes Quarry Park." *Ohio Geology* 2: 7.

13. Clifton Gorge, Greene County

Hansen, M. C. 1998. "The Geology of Ohio—The Silurian." *Ohio Geology* spring: 1, 3–6.

Hull, D. N. 1982. *Guide to the Geology along Interstate 75 between Toledo and Cincinnati*. Educational Leaflet 13. Ohio Department of Natural Resources, Division of Geological Survey.

Schumacher, G. A., et al. 2012. *Geology of the Dayton Region in Core and Outcrop—A Workshop and Field Trip for Citizens, Environmental Investigators, Geologists, and Educators*. Open-File Report 2012–1. Ohio Department of Natural Resources, Division of Geological Survey.

14. Circleville Esker, Pickaway County

Goldthwait, R. P. 1959. "Scenes in Ohio during the Last Ice Age." *Ohio Journal of Science* 59 (4): 193–216.

Hansen, M. C. 1997. *The Ice Age in Ohio*. Educational Leaflet 7. Ohio Department of Natural Resources, Division of Geological Survey.

Hull, D. N. 2001. *Sand and Gravel*. GeoFacts 19. Ohio Department of Natural Resources, Division of Geological Survey.

15. Hueston Woods, Preble County

Davis, R. A. 1998. *Cincinnati Fossils: An Elementary Guide to the Ordovician Rocks and Fossils of the Cincinnati, Ohio, Region*. Cincinnati Museum of Natural History.

Fenneman, N. M. 1916. *Geology of Cincinnati and Vicinity*. Bulletin 19. Geological Survey of Ohio.

Hansen, M. C. 1997. "The Geology of Ohio—The Ordovician." *Ohio Geology* fall: 1–6.

16. Miami Bluffs, Butler County

Ekberg, M. P., et al. 1993. "Late Wisconsinan Glacial Advance and Retreat Patterns in Southwestern Ohio, USA." *Boreas* 22 (3): 189–204.

Martin, W. D. 1975. "The Petrology of a Composite Vertical Section of the Cincinnatian Series Limestone (Upper Ordovician) of Southwestern Ohio, Southeastern Indiana, and Northern Kentucky." *Journal of Sedimentary Petrology* 45 (4): 907–25.

Monroe, J. S., and R. Wicander. 1997. *The Changing Earth: Exploring Geology and Evolution*. West/Wadsworth Publishing Co.

17. Caesar Creek, Warren County

Eckert, A. W. 1961. "The Mammoth Trilobite of Dayton." *Science Digest* July: 67–70.

Shrake, D. L. 1992. *Excursion to Caesar Creek State Park in Warren County, Ohio: A Classic Upper Ordovician Fossil-Collecting Locality*. Guidebook 12. Ohio Department of Natural Resources, Division of Geological Survey.

———. 2005. Isotelus: *Ohio's State Fossil*. GeoFacts 6. Ohio Department of Natural Resources, Division of Geological Survey.

Wells, J. W. 1942. "Supposed Color-Markings in Ordovician Trilobites from Ohio." *American Journal of Science* 240 (10): 710–13.

18. Trammel Fossil Park, Hamilton County

Davis, R. A. 1998. *Cincinnati Fossils: An Elementary Guide to the Ordovician Rocks and Fossils of the Cincinnati, Ohio, Region*. Cincinnati Museum of Natural History.

Haneberg W. C., et al. 1992. *Cincinnati's Geologic Environment: A Trip for Secondary-School Science Teachers*. Guidebook 9. Ohio Department of Natural Resources, Division of Geological Survey.

Hansen, M. C. 1997. "The Geology of Ohio—The Cambrian." *Ohio Geology* winter: 1, 3–5.

19. Rocky Fork Gorge, Highland County

Snyder, T. A. 2009. *Rainbows of Rock, Tables of Stone: The Natural Arches and Pillars of Ohio*. McDonald and Woodward Publishing.

White, G. W. 1926. "The Limestone Caves and Caverns of Ohio." *Ohio Journal of Science* 26 (2): 73–116.

20. Serpent Mound, Adams County

Baranoski, M. T., et al. 2003. *Subsurface Geology of the Serpent Mound Disturbance, Adams, Highland, and Pike Counties, Ohio*. Report of Investigations 146. Ohio Department of Natural Resources, Division of Geological Survey.

Carlton, R. W., et al. 1998. "Discovery of Microscopic Evidence for Shock Metamorphism at the Serpent Mound Structure, South-Central Ohio: Confirmation of an Origin by Impact." *Earth and Planetary Science Letters* 162: 177–85.

Koucky, F. L., and S. P. Reidel. 1987. "The Serpent Mound Disturbance, South-Central Ohio." In *North-Central Section of the Geological Society of America*, Centennial Field Guide, vol. 3, edited by D. L. Biggs, 431–36.

Schumacher, G. A. 2003. "August F. Foerste: The First Geologist to Map the Serpent Mound Disturbance." *Ohio Geology* 2: 1, 3–5.

LAKE PLAIN

Hansen, M. C. 1997. *The Ice Age in Ohio*. Educational Leaflet 7. Ohio Department of Natural Resources, Division of Geological Survey.

Hough, J. L. 1958. *Geology of the Great Lakes*. University of Illinois Press.

White, G. W. 1982. *Glacial Geology of Northeastern Ohio*. Bulletin 68. Ohio Department of Natural Resources, Division of Geological Survey.

21. Headlands Dunes, Lake County

Clifford, M. J. 1973. *Silurian Rock Salt of Ohio*. Report of Investigations 90. Ohio Department of Natural Resources, Division of Geological Survey.

Hansen, M. C. 1983. "Ohio's Salt Industry." *Ohio Geology Newsletter* fall: 1–6.

Kurz, H. 1942. *Florida Dunes and Scrub, Vegetation and Geology*. Geological Bulletin 23. Florida State Geological Survey, 15–146.

Marshall, H. G. 1955. "Succession in a Dune Community at Mentor Headlands, Ohio." *Ohio Journal of Science* 55 (2): 90–94.

22. Rocky River Valley, Cuyahoga County

Anderson, P. S. L., and M. W. Westneat. 2007. "Feeding Mechanics and Bite Force Modeling of the Skull of *Dunkleosteus terrelli*, an Ancient Apex Predator." *Biology Letters* 3 (1): 77–80.

Maisey, J. G. 1996. *Discovering Fossil Fishes*. Westview Press.

23. Kelleys Island Grooves, Erie County

Feldmann, R. M., and T. W. Bjerstedt. 1987. "Kelleys Island: Giant Glacial Grooves and Devonian Shelf Carbonates in North-Central Ohio." In *North-Central Section of the Geological Society of America*, Centennial Field Guide, vol. 3, edited by D. L. Biggs, 95–98.

Forsyth, J. L. 1971. *Geology of the Lake Erie Islands and Adjacent Areas*. Michigan Basin Geological Society, Annual Field Excursion.

Hansen, M. C. 1988. "Glacial Grooves: 'Rock-Scorings of the Great Ice Invasions': Revisited." *Ohio Geology Newsletter* spring: 1–5.

Snow, R. S., et al. 1991. "A Field Guide: The Kelleys Island Glacial Grooves, Subglacial Erosion Features on the Marblehead Peninsula, Carbonate Petrology, and Associated Paleontology." *Ohio Journal of Science* 91 (1): 16–26.

24. Crystal Cave, Ottawa County

Kraus, E. H. 1905. "On the Origin of the Put-in-Bay Caves." *American Geologist* 35: 167–71.

Verber, J. L., and D. H. Stansbery. 1953. "Caves in the Lake Erie Islands." *Ohio Journal of Science* 53 (6): 358–62.

White, G. W. 1926. "The Limestone Caves and Caverns of Ohio." *Ohio Journal of Science* 26 (2): 73–116.

25. Sylvania Fossil Park, Lucas County

Camp, M. J., and C. B. Hatfield. 1991. "Middle Devonian (Givetian) Silica Formation of Northwest Ohio—Description and Road Log." *Ohio Journal of Science* 91 (1): 27–34.

Feldmann, R. M. 2005. *Fossils of Ohio*. Bulletin 70. Ohio Department of Natural Resources, Division of Geological Survey.

Hansen, M. C. 1999. "The Geology of Ohio—The Devonian." *Ohio Geology* 1: 1, 3–7.

26. Maumee River Ledges, Lucas County

Feldmann, R. M. 2005. *Fossils of Ohio*. Bulletin 70. Ohio Department of Natural Resources, Division of Geological Survey.

Hansen, M. C. 1998. "The Geology of Ohio—The Silurian." *Ohio Geology* spring: 1, 3–6.

Janssens, A. 1977. *Silurian Rocks in the Subsurface of Northwestern Ohio*. Report of Investigations 100. Ohio Department of Natural Resources, Division of Geological Survey.

27. Bowling Green Fault, Lucas County

Culotta, R. C., et al. 1990. "A Tale of Two Sutures: COCORP's Deep Seismic Surveys of the Grenville Province in the Eastern US Midcontinent." *Geology* 18 (7): 646–49.

Dean, S. L., et al. 1991. "Field Guide to Joint Patterns and Geomorphological Features of Northern Ohio." *Ohio Journal of Science* 91 (1): 2–15.

Hansen, M. C. 1996. "The Geology of Ohio—The Precambrian." *Ohio Geology* winter: 1–6.

Norris, S. E. 1975. "Geologic Structure of Near-Surface Rocks in Western Ohio." *Ohio Journal of Science* 75 (5): 225–28.

28. Lake Maumee Ridges, Williams County

Farnsworth, D. W. 1958. "Geology along the Maumee River." *Ohio Conservation Bullletin* 22 (1): 16–17, 23.

Hansen, M. C. 1989. "The History of Lake Erie." *Ohio Geology Newsletter* fall: 1, 3–6.

Hough, J. L. 1958. *Geology of the Great Lakes*. University of Illinois Press.

Klotz, J. A., and J. L. Forsyth. 1993. "Late Glacial Origin of the Maumee Valley Terraces, Northwestern Ohio." *Ohio Journal of Science* 93 (5): 126–33.

APPALACHIAN PLATEAU

Horvath, A. 1986. *Hills of Hocking: Southeastern Ohio's Beautiful Appalachian Plateau Area—Photographic Essays*. Allan L. Horvath Publications.

Wallace, W. W. 1940. *The Physiographic Provinces of North America*. Ginn and Company.

29. Nelson Ledges, Portage County

Pettit, L. 1958. "The Effects of Weathering and Other Changes at Nelson Ledges State Park." *Ohio Journal of Science* 58 (3): 182–86.

Ver Steeg, K. 1932. "Nelson Ledge State Park." *Ohio Journal of Science* 32 (3): 177–93.

Williams, A. B. 1940. *Geology of the Cleveland Region*. Cleveland Museum of Natural History.

30. Ritchie Ledges, Summit County

Bartrum, J. A. 1936. "Honeycomb Weathering of Rocks near the Shoreline." *New Zealand Journal of Science and Technology* 18: 593–600.

Blackwelder, E. 1929. "Cavernous Rock Surfaces of the Desert." *American Journal of Science* 17: 393–99.

Mustoe, G. E. 1982. "The Origin of Honeycomb Weathering." *Geological Society of America Bulletin* 93: 108–15.

Prebble, M. M. 1967. "Cavernous Weathering in the Taylor Dry Valley, Victoria Land, Antarctica." *Nature* 216: 1194–95.

31. Clearfork Gorge, Ashland County

Van Buskirk, K. 1983. "The Geology of Mohican State Park." *Ohio Geology Newsletter* summer: 4–6.

Ver Steeg, K. 1947. "Black Hand Sandstone and Conglomerate in Ohio." *Geological Society of America Bulletin* 58 (8): 703–28.

32. Massillon Buffstone, Tuscarawas County

Bownocker, J. A. 1923. "Iron Molding Sands in Ohio." *Ohio Journal of Science* 23 (1): 25–40.

Gray, H. H. 1956. "Petrology of the Massillon Sandstone at the Type Locality." *Ohio Journal of Science* 56 (3): 138–46.

Van Buskirk, K. 1982. "Ohio's Sandstone Industry." *Ohio Geology Newsletter* spring: 1–3.

33. Cherry Valley Coke, Columbiana County

Camp, M. J. 2006. *Roadside Geology of Ohio*. Mountain Press Publishing Company.

Carey, M., and I. Adams. 2007. *Backroads of Ohio: Your Guide to Ohio's Most Scenic Backroad Adventures*. Voyageur Press.

Village of Leetonia. No date. "Cherry Valley Bee-Hive Coke Ovens." www.leetonia.org/cokeovens.asp.

34. Kittanning Clay, Columbiana County

Blankenbeker, S. D. 1999. "The Paving Brick Industry in Ohio." In *1998 Report on Ohio Mineral Industries*, Ohio Department of Natural Resources, Division of Geological Survey, edited by M. Wolfe, 1–11.

Lamborn, R. E., et al. 1938. *Shales and Surface Clays of Ohio*. Bulletin 39. Ohio Department of Natural Resources, Division of Geological Survey.

Stout, W., et al. 1923. *Coal Formation Clays of Ohio*. Bulletin 26. Ohio Department of Natural Resources, Division of Geological Survey.

Wolfe M. E., and S. D. Blankenbeker. 2005. *Clay and Shale in Ohio*. Educational Leaflet 12, rev. ed. Ohio Department of Natural Resources, Division of Geological Survey.

35. Freeport Pulpstone, Jefferson County

Bowles, O. 1917. *Sandstone Quarrying the United States*. US Bureau of Mines Bulletin 124.

Bownocker, J. A. 1915. *Building Stones of Ohio*. Bulletin 18. Ohio Division of Geological Survey.

Crowell, D. L. 1996. "From Pulpstones to Bats." *Ohio Geology* summer: 1, 3–5.

Lamborn, R. E. 1930. *Geology of Jefferson County*. Bulletin 35. Ohio Department of Natural Resources, Division of Geological Survey.

36. Puskarich Coal Museum, Harrison County

Crowell, D. L. 1995. *History of the Coal-Mining Industry in Ohio*. Bulletin 72. Ohio Department of Natural Resources, Division of Geological Survey.

———. 2000. *Coal*. Educational Leaflet 8. Ohio Department of Natural Resources, Division of Geological Survey.

Hansen, M. C. 1990. *Coal: How It Is Found and Used*. Enslow Publishers.

Hook, R. W., and T. R. Miller. 1996. "An Underground Forest." *Ohio Geology* spring: 1, 3.

37. Milepost-50 Cyclothem, Guernsey County

Condit, D. D. 1912. *Conemaugh Formations in Ohio*. Bulletin 17. Geological Survey of Ohio.

Hansen, M. C. 1993. *Guide to the Geology along Interstate 77 between Marietta and Cleveland*. Educational Leaflet 15. Ohio Department of Natural Resources, Division of Geological Survey.

Wanless, H. R., and J. M. Weller. 1932. "Correlation and Extent of Pennsylvanian Cyclothems." *Geological Society of America Bulletin* 43 (4): 1003–16.

38. Blackhand Gorge, Licking County

Hansen, M. C. 2001. "The Geology of Ohio—The Mississippian." *Ohio Geology* 2: 1, 3–7.

Hyde, J. E. 1953. *Mississippian Formations of Central and Southern Ohio*. Bulletin 51. Ohio Department of Natural Resources, Division of Geological Survey.

Keirns, A. J. 1995. *Blackhand Gorge: A Journey through Time*. Little River Publishing.

39. Flint Ridge, Licking County

Carlson, E. H. 1987. "Flint Ridge, Ohio: Flint Facies of the Pennsylvanian Vanport Limestone." In *North-Central Section of the Geological Society of America*, Centennial Field Guide, vol. 3, edited by D. L. Biggs, 415–18.

Getz, G. L. 2012. *Flint: Ohio's Official Gemstone*. Educational Leaflet 6. Ohio Department of Natural Resources, Division of Geological Survey.

Mills, W. C. 1921. "Flint Ridge." *Ohio Archaeological and Historical Quarterly* 30: 91–161.

Stout, W., and R. A. Schoenlaub. 1945. *The Occurrence of Flint in Ohio*. Bulletin 46. Ohio Department of Natural Resources, Division of Geological Survey.

40. Ames Marker Bed, Noble County

Camp, M. J. 2006. *Roadside Geology of Ohio*. Mountain Press Publishing Company.

Hansen, M. C. 1993. *Guide to the Geology along Interstate 77 between Marietta and Cleveland*. Educational Leaflet 15. Ohio Department of Natural Resources, Division of Geological Survey.

41. Thorla-McKee Well, Noble County

Bownocker, J. A. 1903. *The Occurrence and Exploitation of Petroleum and Natural Gas in Ohio*. Bulletin 1. Ohio Department of Natural Resources, Division of Geological Survey.

Hansen. M. C. 1993. "The First Oil Well." *Ohio Geology* spring: 1–4.

Spencer, J. A., and M. J. Camp. 2008. *Ohio Oil and Gas*. Images of America. Arcadia Publishing.

Spencer, J. A., and J. Robinson. 2007. "The Thorla-McKee Salt Works and Oil Wells and the Macksburg, Ohio, Oil Booms." *Oil-Industry History* 7 (1): 83–91.

42. Rockbridge, Hocking County

Hall, J. F. 1961. *The Geology of Hocking State Park*. Information Circular 8. Ohio Department of Natural Resources, Division of Geological Survey.

Hansen, M. C. 1975. *Geology of the Hocking Hills State Park Region*. Guidebook 4. Ohio Department of Natural Resources, Division of Geological Survey.

Snyder, T. A. 2009. *Rainbows of Rock, Tables of Stone: The Natural Arches and Pillars of Ohio*. McDonald and Woodward Publishing.

———. 2010. *Ohio's Natural Arches*. Educational Leaflet 18. Ohio Department of Natural Resources, Division of Geological Survey.

43. Rock House, Hocking County

Carman, J. E. 1946. "The Geologic Interpretation of Scenic Features in Ohio." *Ohio Journal of Science* 46 (5): 241–83.

Hall, J. F. 1961. *The Geology of Hocking State Park*. Information Circular 8. Ohio Department of Natural Resources, Division of Geological Survey.

Hansen, M. C. 1975. *Geology of the Hocking Hills State Park Region*. Guidebook 4. Ohio Department of Natural Resources, Division of Geological Survey.

Snyder, T. A. 2009. *Rainbows of Rock, Tables of Stone: The Natural Arches and Pillars of Ohio*. McDonald and Woodward Publishing.

44. Old Mans Cave, Hocking County

Hansen, M. C. 1975. *Geology of the Hocking Hills State Park Region*. Guidebook 4. Ohio Department of Natural Resources, Division of Geological Survey.

Hyde, J. E., and M. F. Marple. 1953. *The Mississippian Formations of Central and Southern Ohio*. Bulletin 51. Ohio Department of Natural History, Division of Geological Survey.

Maniskas, J. S. 2011. *Logan and Hocking County*. Images of America. Arcadia Publishing.

45. Appalachian Escarpment, Ross County

Fenneman, N. M. 1938. *Physiography of Eastern United States*. McGraw-Hill Book Company.

Price, P. H. 1931. "The Appalachian Structural Front." *Journal of Geology* 39 (1): 24–44.

46. Massieville Terminal Moraine, Ross County

Camp, M. J. 2006. *Roadside Geology of Ohio*. Mountain Press Publishing Company.

Roy, M., et al. 2004. "Glacial Stratigraphy and Paleomagnetism of Late Cenozoic Deposits of the North-Central United States." *Geological Society of America Bulletin* 116 (1–2): 30–41.

Swinford, E. M. 2004. "What the Glaciers Left Behind—The Drift-Thickness Map of Ohio." *Ohio Geology* 1: 1, 3–5.

47. Copperas Mountain, Ross County

Carlson, E. H. 1987. "Copperas Mountain: Black Shale Facies of the Upper Devonian Ohio Shale in South-Central Ohio." In *North-Central Section of the Geological Society of America*, Centennial Field Guide, vol. 3, edited by D. L. Biggs, 427–30.

Clifton, H. E. 1957. "The Carbonate Concretions of the Ohio Shale." *Ohio Journal of Science* 57 (2): 114–24.

Hansen, M. C. 1994. *Ohio Shale Concretions*. GeoFacts 4. Ohio Department of Natural Resources, Division of Geological Survey.

48. Buckeye Furnace, Jackson County

Collins, H. R. 1986. "The Hanging Rock Iron Region of Ohio." *Ohio Geology Newsletter* winter: 1–5.

Keeler, V. D. 1933. "An Economic History of the Jackson County Iron Industry." *Ohio Archaeological and Historical Quarterly* 42 (2): 132–238.

Stout, W. 1933. "The Charcoal Iron Industry of the Hanging Rock Iron District—Its Influence on the Early Development of the Ohio Valley." *Ohio Archaeological and Historical Quarterly* 42 (1): 72–104.

49. Scioto Saline, Jackson County

Bownocker, J. A. 1906. *Salt Deposits and the Salt Industry in Ohio*. Bulletin 8. Geological Survey of Ohio.

Clifford, M. J. 1973. *Silurian Rock Salt of Ohio*. Report of Investigations 90. Ohio Department of Natural Resources, Division of Geological Survey.

Hansen, M. C. 1983. "Ohio's Salt Industry." *Ohio Geology Newsletter* fall: 1–6.

Heimlich, R. A., et al. 1974. "General Geology of the International Salt Company Cleveland Mine, Cleveland, Ohio." In *Selected Field Trips in Northeastern Ohio*, Guidebook 2, Ohio Department of Natural Resources, Division of Geological Survey, edited by R. A. Heimlich and R. M. Feldmann.

50. Teays River Valley, Scioto County

Coffey, G. N. 1961. "Major Preglacial, Nebraskan, and Kansan Glacial Advances in Ohio, Indiana, and Illinois." *Ohio Journal of Science* 61: 295–313.

Hansen, M. C. 1987. "The Teays River." *Ohio Geology Newsletter* summer: 1–6.

Melhorn, W. N, and J. P. Kempton, eds. 1991. *Geology and Hydrogeology of the Teays-Mahomet Bedrock Valley System*. Geological Society of America Special Paper 258.

Ver Steeg, K. 1946. "The Teays River." *Ohio Journal of Science* 46 (6): 297–307.

INDEX

Page numbers in bold indicate an illustration or information in a caption.

Born in Ohio, **Albert B. Dickas** earned BA and MA degrees from Miami University (Oxford, OH). After serving in the US Navy, he earned a PhD at Michigan State University and then worked with Magnolia Petroleum (Mobil Oil) in the offshore salt dome province of Louisiana. From there he moved to Standard Oil of California, where he specialized in hydrocarbon exploration in the subsurface submarine canyons of the Sacramento Valley. Having satisfied his strong belief that firsthand industrial experience is a prerequisite for work in academia, he joined the faculty of the University of Wisconsin–Superior, where he taught for thirty-one years. He also founded an environmental research center and became intensely involved in the industrial exploration for oil in the Precambrian strata of the Midcontinent Rift System. During this time he led numerous field conferences, authored and coauthored more than fifty papers, and delivered presentations from Nova Scotia to Siberia—all on the subject of Precambrian oil and Precambrian rifting. Today he lives on the crest of Brush Mountain in southwest Virginia, where he continues to engage in research and plan travel excursions in his quest for new and interesting geological sites on all seven continents. He also wrote *101 American Geo-Sites You've Gotta See*, also published by Mountain Press (2012).